EINSTEIN FOR BEGINNERS

$E=MC^2$

EINSTEIN

FOR BEGINNERS

JOSEPH SCHWARTZ & MICHAEL McGUINNESS

Pantheon Books   New York

Library of Congress Cataloging-in-Publication Data
Schwartz, Joseph, 1938–
Einstein for beginners.
Bibliography: p.
1. Einstein, Albert, 1879-1955. I. McGuinness, Michael,
1935-   , joint author. II. Title.
QC16.E5S32   530.1'1   79-1889
ISBN 0-375-71459-6

www.pantheonbooks.com
Printed in the United States of America

6 8 ['03] 9 7 5

About the Author and Illustrator

Joseph Schwartz, who is Associate Professor of Physics at the City University of New York, received his Ph.D. in higher energy physics from the University of California in 1964. His many scientific articles have appeared in *Nature, New Scientist,* and other magazines.

Michael McGuinness studied fine arts at the Royal Academy in London. He is a former art director at *Reader's Digest* and designer for the *Observer.*

'If relativity is proved right the Germans will call me a German, the Swiss will call me a Swiss citizen, and the French will call me a great scientist.

If relativity is proved wrong the French will call me a Swiss, the Swiss will call me a German, and the Germans will call me a Jew.'

Albert Einstein was born in Ulm, Germany on March 14, 1879 into a world not of his own making.

Just like the rest of us.

5

What was going on in the world?

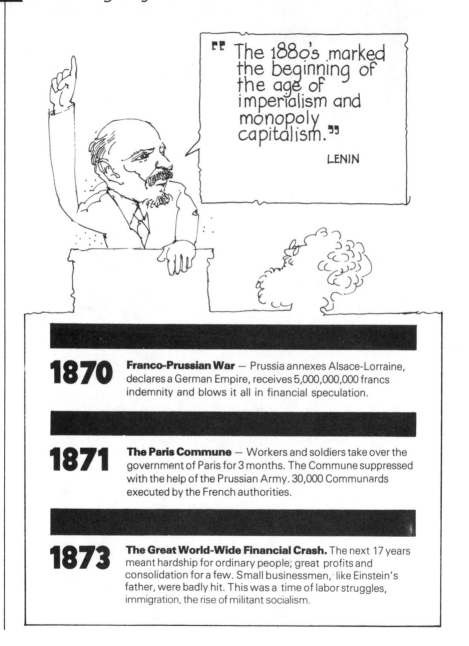

" The 1880's marked the beginning of the age of imperialism and monopoly capitalism. "

LENIN

1870 **Franco-Prussian War** — Prussia annexes Alsace-Lorraine, declares a German Empire, receives 5,000,000,000 francs indemnity and blows it all in financial speculation.

1871 **The Paris Commune** — Workers and soldiers take over the government of Paris for 3 months. The Commune suppressed with the help of the Prussian Army. 30,000 Communards executed by the French authorities.

1873 **The Great World-Wide Financial Crash.** The next 17 years meant hardship for ordinary people; great profits and consolidation for a few. Small businessmen, like Einstein's father, were badly hit. This was a time of labor struggles, immigration, the rise of militant socialism.

1878

Bismarck passes anti-socialist laws to suppress working-class political agitation.

" The great questions of the day will not be settled by resolutions and majority votes but by blood and iron. "

Otto Von Bismarck
1815-1898

Chancellor of Germany
1871 – 1890

1879

Wilhelm Marr coins the word anti-Semitism and founds the **League of Anti-Semites.**

Jews get the blame for the financial crisis.

7

8

It's a period of tremendous overall industrial expansion. People throughout Europe are forced off the land and into the cities.

The rural Jewish population of
southern Germany falls by 70%
between 1870 and 1900.
Many emigrate to the Americas.

In 1880 Albert's father's business fails because of the depression and the family moves from Ulm, population 1,500, to Munich, population 230,000. Albert is one year old.

> Pauline, I think things are better in Munich.

> Fine! You can go into business with your brother there.

Hermann Einstein
1847 - 1902
Albert's father.
Freeman of Buchau.
Jews were not completely
emancipated until 1867, so
being a freeman was special.

Pauline Koch
1858 - 1920
Albert's mother.
Daughter of a court
purveyor, Julius Koch-
Bernheimer.

Central to Germany's industrialization is the growth of the chemical and electrical industries.

The heavy chemical industry:
bulk production of soda, nitrates, soap and sulphuric acid for bleaching, dyeing, printing, explosives and fertilizers.

Soap

Light chemical industry:
aniline dyes, pharmaceutical products, plastics.

Intensive scientific research into the possibility of making use of coal by-products.

Formation of cartels:
I. G. Farben, Krupp, etc.

Employ-ment for professors

PURE MEDICINES
CHEMICALS
PERFUMERIES
DRUGS
PAINTS, OILS
&C.

Good for
further
experiments.

Makes
profits.

Signaling by Electricity 1837:
telegraphs, cables, batteries, terminals, insulated
wire coils, switches, measuring instruments.

Electroplating 1840:
for fancy tableware and household objects for
the prosperous middle classes.

Electric Lighting 1860-80:
arc lighting for streets, docks, railways and finally
homes.

Electric Power Production 1880:
electrification of railways, furnaces, machinery,
construction of power plants and distribution
systems.

1881. In the suburbs of Munich, Albert's father opens a small factory with his brother Jacob, a trained engineer. They manufacture dynamos, electric instruments and electric arc lights.

Hermann and Jacob are part of the German electrical industry ○○○○○○○

which is going through a period of intense **monopolization**.

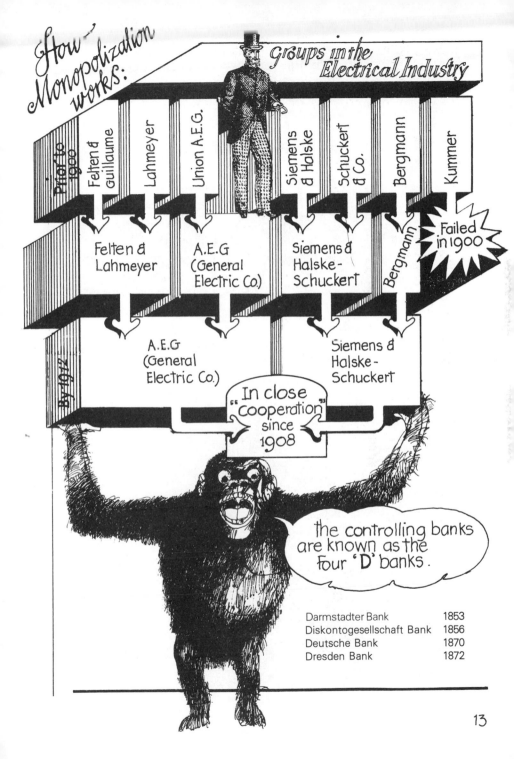

By 1913, half the world's trade in electro-chemical products was in German hands.

Who had the other half?

Glad you asked.

The US of A. General Electric Co., a combine of Thomson-Houston & Edison Co.

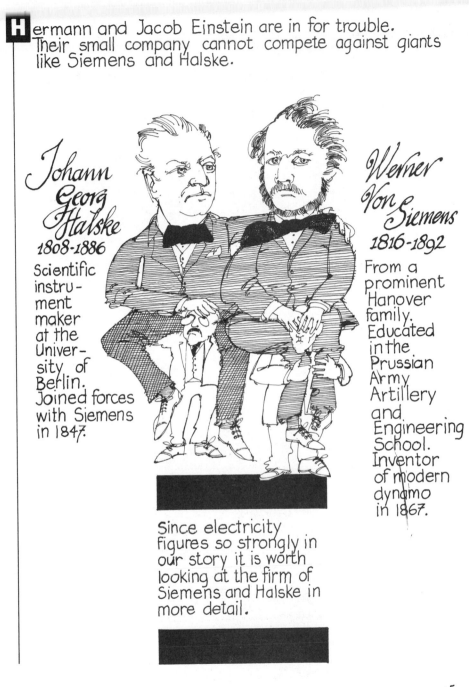

Hermann and Jacob Einstein are in for trouble. Their small company cannot compete against giants like Siemens and Halske.

Johann Georg Halske
1808-1886

Scientific instrument maker at the University of Berlin. Joined forces with Siemens in 1847.

Werner Von Siemens
1816-1892

From a prominent Hanover family. Educated in the Prussian Army Artillery and Engineering School. Inventor of modern dynamo in 1867.

Since electricity figures so strongly in our story it is worth looking at the firm of Siemens and Halske in more detail.

Siemens' first invention was an improved process for gold and silver plating.

With his brother Charles acting as agent, he sold the rights to Elkington of Birmingham, U.K., in 1843.

Siemens joins the circle of Berlin University scientists. He develops an improved telegraph system. This is a method of covering the wire with seamless insulation made of cheap material (gutta-percha: a rubberlike plant substance.)

In 1847 he founds **Telegraphen Bauenstadt von Siemens und Halske** to manufacture and instal telegraph systems.

In 1848 he gets the Prussian government contract to build a network in Northern Germany.

Siemens loses the Prussian contract in 1850. But in Russia he succeeds in selling the Tsar on an extensive system.

The first transatlantic
cable is laid
between 1857 - 1868.

Siemens organizes the Indo - European telegraph in 1870.
It connects London - Berlin - Odessa - Teheran and
Calcutta. He becomes consultant to the British government.
His ship, the Faraday, lays 5 transatlantic
cables between 1875 - 1885.

Electric power becomes a commodity.
The first market is lighting for docks, railways and streets.

In the US it's Thomas Edison who Switches on.

One happy family!

Schuckert, who combines with Siemens, worked with Edison in New Jersey.

Edison organizes the construction of the first central generating station in 1882.

This should turn a nice profit, hey?

Pearl St. Station of
Edison Electric Illuminating Company

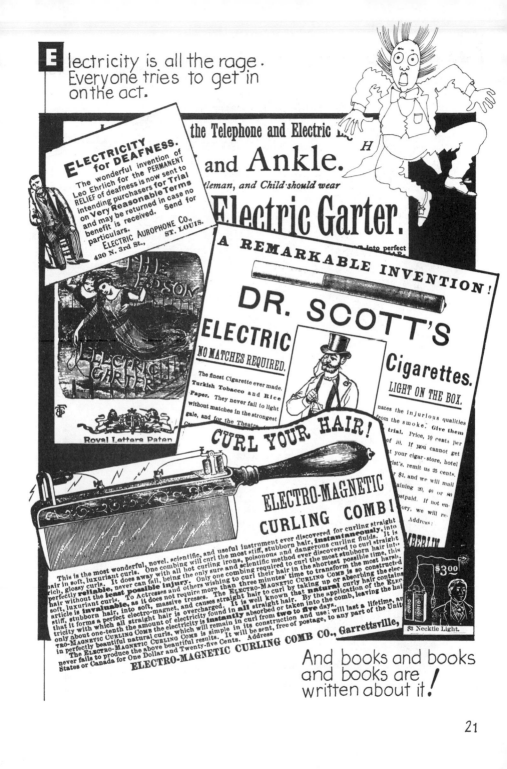

Electricity is all the rage. Everyone tries to get in on the act.

And books and books and books are written about it!

In 1887 the German government opens the **Physikalische - Technische - Reichsanstalt** for research in the exact sciences and precision technology. Siemens donates 500,000 marks to the project. His old friend, Hermann von Helmholtz of the University of Berlin circle, is appointed head.

... and my daughter married Siemens' son.

Hermann Von Helmholtz
1821-1894

Educated as an army surgeon. As the new professor of natural philosophy in Berlin in 1871 he initiated systematic tests of all existing theories of electricity and magnetism. His lab becomes the center of physics research on the Continent. He is known as the patriarch of German science and the state's foremost science advisor.

So Albert was born when electricity had become big business and the most popular of the sciences. His future would be greatly influenced by the commitment of the German state to technical education and state-supported research.

In 1881 Albert's sister Maja is born.

Our family was very close-knit and very hospitable.

Maja Einstein 1881–1951

Albert proves to be a slow, dreamy child. Even at age 9 he spoke hesitantly.

Albert's closest friend in childhood.

Albert's Germany is a very military place . . . .

Arms expenditure nearly *triples* between 1870 and 1890.

The officer corps increases from 3000 to 22,500. Three year military service is compulsory. Socialist literature is forbidden. Youths are subjected to fear and humiliation.

Veterans organizations are state supported. Membership increases from 27,000 in 1873 to 400,000 in 1890 and 1,000,000 in 1900.

Heads of state all appear in military uniform.

Even the taxi drivers wear uniforms.

Albert doesn't like it.

Wilhelm I
Emperor of Germany
from 1871-1888

25

Albert goes to school...

...which is very military.

Albert goes to a Catholic school.

He is the only Jew in his class....

(Albert's father was a non-religious Jew who regarded the kosher dietary laws as ancient superstition.)

Albert had a much better time at home playing with his sister Maja.

Albert's **uncle** Jacob introduces him to math ○○○○○○○○

I like my uncle Jake. He always shows me things.

Algebra is a merry science. When the animal we are hunting cannot be caught, we call it x temporarily and continue to hunt it until it is caught.

○○○○○○○○ And his mother introduces him to music and literature.

Oh, no, not violin lessons! It's just like school.

Go on, you know you like to play when your cousins come.

It was a Jewish custom in southern Germany to invite a poor Jew to dinner on Thursdays. Max Talmey, a medical student in Munich, visited the Einstein home when Albert was 12.

Great public interest in science in Germany produced popular science best-sellers and vice versa. Talmey brought some of these with him.

BERNSTEIN
Naturwissenschaftliches
Volksbuch

Wow! nice pictures! Thanks Max.

animals plants stars

meteors volcanoes climate

And more significantly Max followed up Uncle Jake's teaching of algebra with a book on geometry

Spieker **Lehrbuch der ebenen Geometrie**

Hmmm, that looks interesting.

With Talmey's assistance, Albert worked through Spieker's Plane Geometry and later went on to teach himself the elements of calculus.

Hermann, do you think Albert reads too much?

Better he should read than do nothing!

A lbert's reading undermines his faith in authority.

Through the reading of popular scientific books I soon reached the conviction that much of the stories in the Bible could not be true. The consequence was a positive fanatic orgy of free thinking coupled with the impression that youth is intentionally being deceived by the State through lies. It was a crushing impression. Suspicion against every kind of authority grew out of this experience, a skeptical attitude towards the convictions which were alive in any specific social environment — an attitude which never left me, even though later on, because of a better insight into causal connections, it lost some of its original poignancy.

In 1894 Hermann's business fails. The family moves south to Milan, Italy.

Albert, you'll stay here to finish school and get your diploma, you'll need it.

Like hell I will.

After two months on his own, Albert obtains a doctor's certificate saying that he is suffering a nervous breakdown. The school authorities dismiss him

Just what I wanted!

Out you go with pleasure.

Papa, I'm renouncing my German citizenship.

I'm off to the mountains. I think I'll visit our cousins in Genoa.

What am I going to do with him? No diploma either. Oy!

Albert spends a free happy year in Italy. But his father's business fails again. The family moves to Pavia where **again** it fails!

Business?! Ha! I will become a theoretical physicist.

Albert, I can no longer support you. You must become an engineer and go into business.

Even when I was a fairly precocious young man, the nothingness of the hopes and stirrings which chases most men restlessly through life came to my consciousness with considerable vitality. Moreover I soon discovered the cruelty of that chase, which in those years was much more carefully covered up by hypocrisy and glittering words than is the case today. By the mere existence of the stomach everyone was condemned to participate in that chase.

Without a diploma, Albert can't enter University. But the **Eidgenossiche Technische Hochschule,** the ETH, in Zurich, the most elite technical school outside of Germany, would admit him if he passed an entrance exam. He failed miserably.

Albert has a good time in Aarau.

Ooh, that Albert Einstein is <u>Cute</u>.

He stays with the headmaster of the school, Professor Winteler, who has a son, Paul, and a daughter Albert's age. Albert's sister Maja later marries Paul Winteler. He studies physics with....

.... August Tuschmid, considered a first-class teacher of physics.

The central problem in physics today is the resolution of Newton's mechanical world view with the new equations of electromagnetism.

Hmmm....

At the end of the year Albert graduates
and passes his ETH exam.

Einstein's graduating class Aarau 1896

On 28 January 1896 Albert's official application for the termination of his German nationality is approved. He becomes a **stateless person!** Albert convinces his father that he should be a teacher instead of an engineer. In October 1896 he is ready for the....
...."big time".

The big time – what's he mean?

Dunno, let's see.

The ETH was a Big League outfit. The Physics Institute was planned by Heinrich Weber and his friend Siemens.

Heinrich Friedrich Weber 1842-1913

...It attracted world-wide attention

Description by Henry Crew, PhD, U.S. physics professor in 1893 :

"H. F. Weber and Dr Pernet are at the head of the physics department in the Polytechnic. They not only have the most complete instrumental outfit I have ever seen, but also the largest building I have ever seen used for a physical laboratory. Tier on tier of storage cells, dozens and dozens of the most expensive tangent and high resistence galvanometers, reading telescopes of the largest and most expensive form by the dozen, 2 or 3 in each room. The apparatus cost 400,000 francs, the building alone 1 million francs."

But the engineers at the ETH complained that their teachers were too abstract.

The students organized demonstrations against the mathematics lectures.

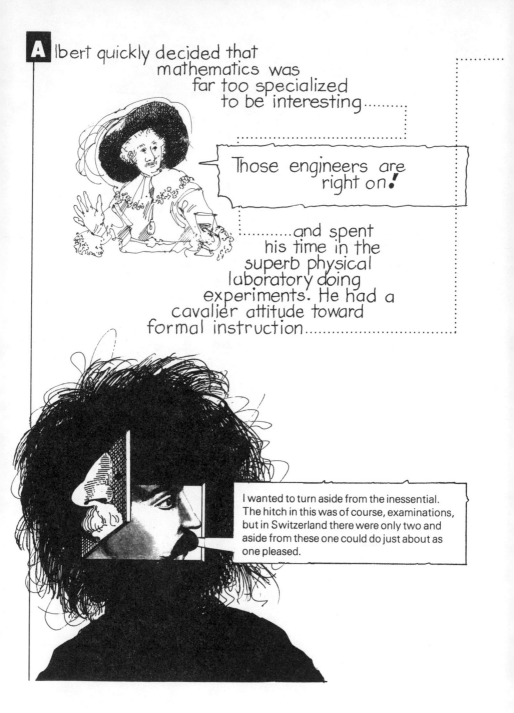

Albert quickly decided that mathematics was far too specialized to be interesting..........

Those engineers are right on!

..........and spent his time in the superb physical laboratory doing experiments. He had a cavalier attitude toward formal instruction..........

I wanted to turn aside from the inessential. The hitch in this was of course, examinations, but in Switzerland there were only two and aside from these one could do just about as one pleased.

...and naturally he quickly antagonized some of his instructors.

You're clever, Einstein, extremely clever. But you have one great fault: you never let yourself be told anything!

Yes, Herr Weber....

He hates it when I don't call him Herr **Professor!**

and Albert

never attended

lectures.

Hey, Marcel, what happened in differential geometry today?

Same old stuff! Here's the notes.

Let's have lunch.

Albert gets 100 francs a month from his relatives. He saves 20 francs of it each month toward his Swiss citizenship.

Expensive.... and restricted to a few applicants.

He forms friendships with Michelangelo Besso, "the finest sounding-board in all Europe," Marcel Grossmann, who later helps Albert get his first secure job in the Swiss Patent Office, and Mileva Maric, a mathematician from Serbia whom he marries in 1903. They have a good time in the lively political atmosphere of Zurich.

Exiled revolutionaries from Germany and Russia all come to Zurich. Alexandra Kollontai, Trotsky, Rosa Luxemburg, and later Lenin, are there.

Trotsky

Rosa Luxemburg

Alexandra Kollontai

Lenin

Albert learns about revolutionary socialism from his friend Friedrich Adler, a junior lecturer in physics.

Friedrich Adler

Friedrich is the son of Victor Adler, the leader of the Austrian Social Democrats, sent by his father to study physics "and forget politics" But Adler remains involved in the socialist movement. In 1918 he assassinates the Austrian Prime Minister. Albert submits testimony on his behalf. Friedrich gets amnesty and doesn't serve any time.

In physics Newton's consolidation of the laws of mechanics had dominated for the previous 200 years.

Isaac Newton
1642-1727

Attended Trinity College, Cambridge. Whig MP for Cambridge 1689-1690. Long–term interest in metallurgy led to his becoming Master of the Mint from 1696 to his death in 1727. Founder of the theoretical basis of mechanics. Using Kepler's summary of the measurements of the motions of the planets he formulated laws of motion of material objects.

Clock-work.

Opticks.

Newton's mechanical world view is part of 18th & 19th century European philosophy and vice versa.

Albert was skeptical but nevertheless impressed by the achievements of the mechanical world view.

Dogmatic rigidity prevailed in all matters of principles. In the beginning God created Newton's law of motion together with the necessary masses and forces.

But what the 19th century achieved on this basis was bound to arouse the admiration of every receptive person.

Albert, like most beginning physics students, particularly admired the ability of mechanics to explain the behavior of gases. The relationship between the pressure, volume and temperature of a gas could be derived by treating the particles of a gas as projectiles constantly bombarding the walls of the container. From this treatment came a number of impressive results: the way the energy of a gas depended on temperature, how viscous a gas is, how well it conducts heat and how fast it can diffuse. Comparison of this model to experiment also yielded the first estimates of the sizes of atoms.

But it was the physics of electricity and the electrodynamics of Faraday, Maxwell and Hertz that most attracted his attention. . . .

Faraday: the most accomplished experimental physicist of the 19th C. Son of a blacksmith.

1

He worked 7 years as a bookbinder before coming to the attention of Sir Humphrey Davy.

2

3 Sir Humphrey Davy was head of the Royal Institution in London. Faraday became Davy's assistant and had to endure the routine insults of the British class system throughout his early years. Davy's wife refused to eat at the same table with him and demanded that Davy do the same.

Michael Faraday
1791-1867

4 In 1832 Faraday published the experimental and theoretical work that paved the way for Maxwell's theory of electromagnetism. His work was hampered in later years by a failure of memory caused by mercury poisoning.

Child of a prominent Edinburgh family. From 1857 to 1864 he worked at putting Faraday's results into mathematical form.

James Clerk Maxwell 1831 - 1879

Maxwell's equations showed that electric and magnetic forces should move through empty space at exactly the speed of light.

Hmm.... Faraday's picture of lines of force traversing all space is a good one. I think I can use that.

I'm Helmholtz

Maxwell expressed himself in obscure and contradictory language so his results weren't accepted in Europe. In 1871, I waded through his papers and realized that he was probably right. I put my best student on the problem of showing experimentally that the electric force propagated at the speed of light.

propagate = spread out; get from place to place.

Helmholtz's best student was . . . .

Son of a lawyer and Senator of Hamburg. Trained as an engineer, he became attracted to Helmholtz's lab in Berlin. In 1886, after 8 years of work on Maxwell's theory, he demonstrated experimentally that the electric force propagates through space at the speed of light.

Heinrich Hertz 1857-1894

WHICH PREPARED THE WAY FOR RADIO

Signor Guglielmo Marconi 1874-1937

Hertz's experiments were widely popularized and inspired the 20-year-old **Guglielmo Marconi.** Working with Professor Augusto Righi, a friend and neighbor in Bologna, Marconi built signaling devices.

I tried to sell the British admiralty a self-propelled torpedo in 1896.

50

Albert got very excited about this line of work.

The incorporation of optics into the theory of electromagnetism with its relation to the speed of light to electrical and magnetic measurements . . . was like a revelation!

Electricty? Magnetism? Optics?

Science is mysterious.

Science is a force in production.

Hey, what about curiosity?

Science is social relations.

Science isn't neutral.

THE MECHANICAL ARTS SIMPLIFIED

How far would Albert's childhood curiosity about the magnet have gotten without a **social basis**? Without the organized **work** of many people like Faraday, Maxwell, Hertz and others?

> Knowledge accumulates through **work**

'Curiosity' is just a way of saying that human beings can **change** their environment, can **improve** things, can **discover** what is useful or not...

The history of electricity and magnetism shows us how this process works going right back to ancient times....

If only we could **use** those volcanoes to warm us in winter!

Natural magnets, or lodestones, were reported by the Chinese circa 2600 B.C.

When you dig for iron, you find lots of them.

Lodestones are magnetized by the Earth's own magnetism. Also called magnetite. It is an oxide of iron (iron combined with oxygen).

The Chinese used them first for burial purposes and only later for navigation.

There were occult specialists in China called geomancers. Their job was to see that a person's grave was correctly lined up for proper entry to the after-life.

Far out!

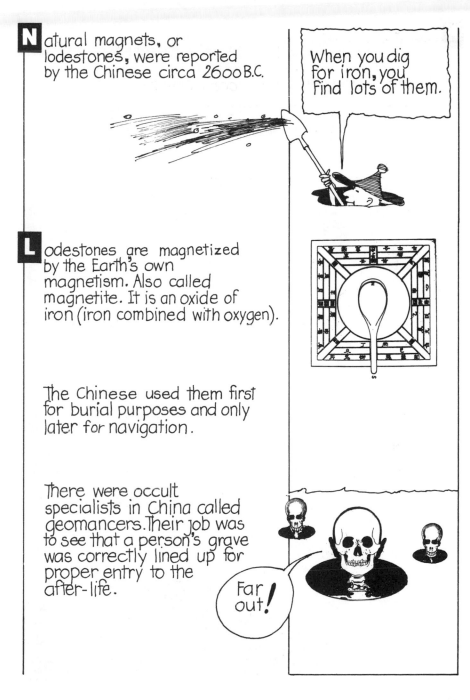

Around 9oo B.C. magnetized needles began to be used as direction indicators.

Lucretius (circa 55 B.C.) wrote a poem about magnetism:

" *The steel will move to seek the stone's embrace Or up or down or t'any other place* "

FROM <u>DE RERUM NATURA</u>

And that was that for 16oo years. Magnetism was good for directions and as a curiosity for the leisured.

"My brother told me that Bathanarius produced a magnet and held it under a silver plate on which he placed a bit of iron. The intervening silver was not affected at all, but, precisely as the magnet was moved backward and forward below it, no matter how quickly, so was the iron attracted above."

from The City of God

Saint Augustine A.D. 354-430

Electricty has a similar history.

The Greeks circa 400 B.C.

knew that rubbing amber attracted bits of straw......

.... Etruscans have a method for controlling lightning.

And that's where **that** stood for a very long time!

In 1726 a student of Newton's, Stephen Gray, showed that frictional electricity....

.... can be made to travel along a hemp thread.

By the end of the 18thC. a number of people like Coulomb in France and Galvani and Volta in Italy, supported by wealthy patrons, were exploring the phenomena of electricity.

Volta invented a battery which made steady currents available for the first time.

Interest dropped off in frictional electricity and everyone rushed to make batteries because they were so much **better.**

Coulomb made detailed measurements of the electric force. His experiments showed that a formula could be written for the electric force similar to Newton's formula for gravitation.

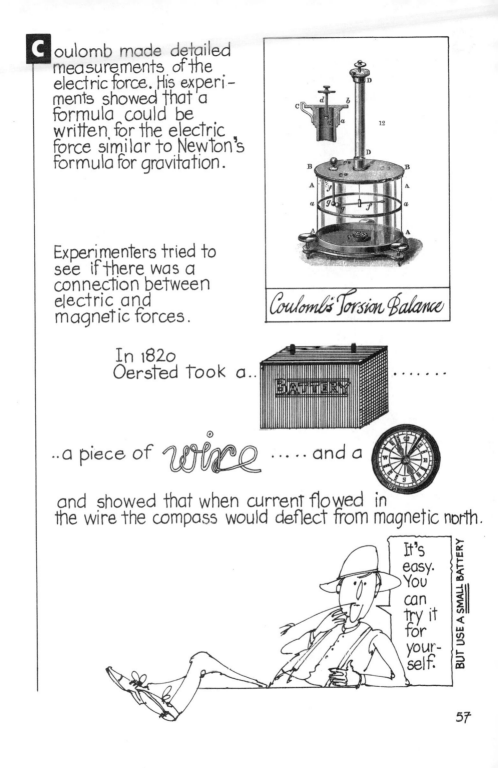

Coulomb's Torsion Balance

Experimenters tried to see if there was a connection between electric and magnetic forces.

In 1820 Oersted took a.. . . . .

..a piece of *wire* . . . . . and a

and showed that when current flowed in the wire the compass would deflect from magnetic north.

It's easy. You can try it for yourself.

BUT USE A SMALL BATTERY

57

André Ampère made even more precise measurements of this new force exerted by currents flowing through wires.

Ampère's discovery was elegant but Oersted's was **commercial.** Electric telegraphy became possible because the electric current could be used to deflect a magnetized needle somewhere else and hence pass on messages!

Having shown that electricity in the form of electric current could produce magnetic effects, it now remained to be shown that magnetism could produce **electric** effects.

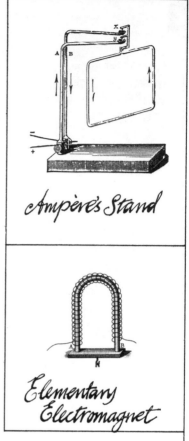

Ampère's Stand

Elementary Electromagnet

This proved to be a tough nut which was not cracked until 1831 by Faraday.

Faraday was able to show finally that you could get an electric current from magnetism.
(The magnetism had to change. A static magnetic force couldn't do it.)

It had been a big gamble and a lot of hard work.

> October 3. 1831
>
> Dear Richard,
>
> I am busy just now again on electromagnetism, and I think I may have got hold of a good thing, but can't say. It may be a weed instead of a fish that, after all my labour, I may at last pull up.
>
> Yours Michael.

This discovery showed that you could get an electric current from the mechanical motion of magnets.

Most everyone dropped research into batteries and started building generators.

Hippolyte Pixii's was the first···   ····

....which was a long way from Siemens' first dynamo in 1867.

And at the same time people started experimenting with electric motors....

....which didn't pay off until wide-scale distribution of power became profitable in the 1880's.

But the key thing for our story is how Faraday tried to understand the effect he observed.

Faraday was one of the very few working-class scientists. His background of rich practical experience served him well in his experimental work. And his overall perspective was very down to earth.

Instead of trying to make up elegant force 'laws', Faraday tried to visualize what was happening when a magnet and a current interacted. So he made **pictures** of what was happening.

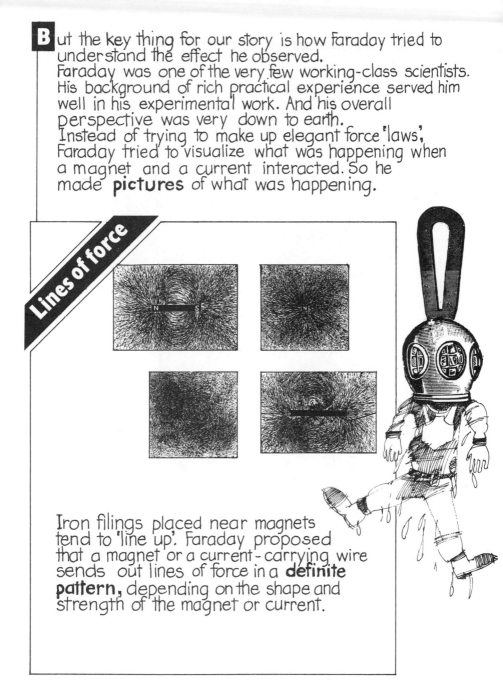

Lines of force

Iron filings placed near magnets tend to 'line up'. Faraday proposed that a magnet or a current-carrying wire sends out lines of force in a **definite pattern,** depending on the shape and strength of the magnet or current.

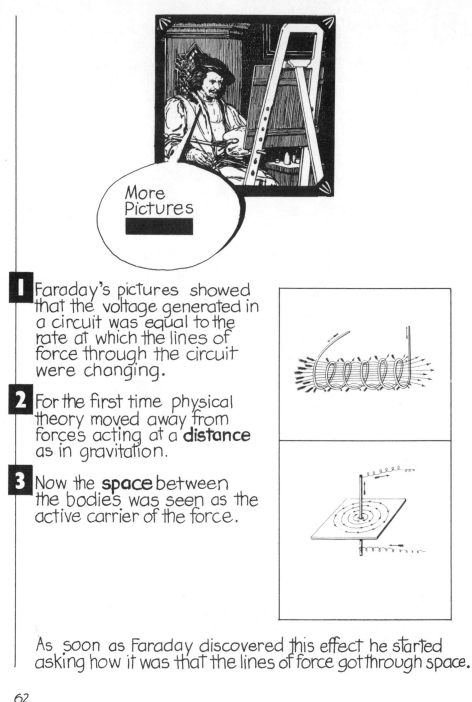

More
Pictures

1 Faraday's pictures showed that the voltage generated in a circuit was equal to the rate at which the lines of force through the circuit were changing.

2 For the first time physical theory moved away from forces acting at a **distance** as in gravitation.

3 Now the **space** between the bodies was seen as the active carrier of the force.

As soon as Faraday discovered this effect he started asking how it was that the lines of force got through space.

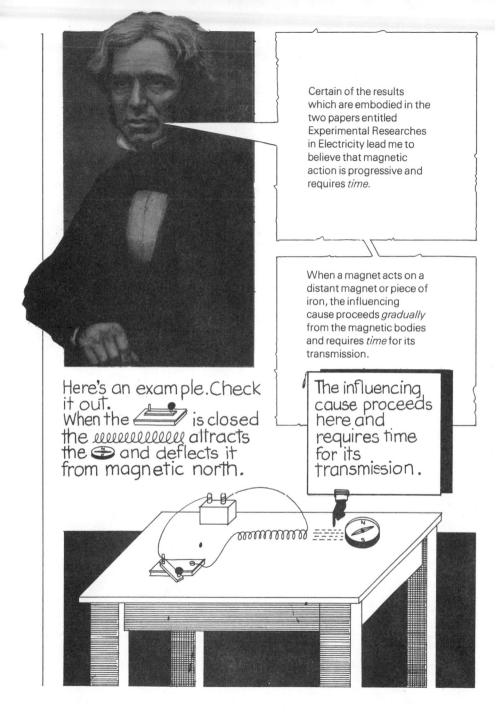

Certain of the results which are embodied in the two papers entitled Experimental Researches in Electricity lead me to believe that magnetic action is progressive and requires *time*.

When a magnet acts on a distant magnet or piece of iron, the influencing cause proceeds *gradually* from the magnetic bodies and requires *time* for its transmission.

Here's an example. Check it out. When the ⬜ is closed the ⚏⚏⚏⚏⚏⚏ attracts the ⊖ and deflects it from magnetic north.

The influencing cause proceeds here and requires time for its transmission.

25 years later Maxwell made very good use of this picture. He renamed the magnetic lines of force the *magnetic field*. He renamed the electric lines of force the *electric field*. He produced equations showing how the fields were related to each other. And, as an extra bonus, the equations predicted that under certain conditions the fields (lines of force, magnetic influence, it's all the same) should move like *waves* through space at the speed of light.

Impressive

Complicated mathematics though.

speed of light?

Measurements of the speed of light

1670	I. Newton	instantaneous
1676	O. Roemer	141,000 miles / sec
1727	J. Bradley	186,233 miles/sec
1849	H. Fizeau	194,000 mi/sec
1875	A. Cornu	186,400 mi/sec
1926	A. Michelson	186,281 mi/sec
1941	C. D. Anderson	186,269 mi/sec
modern value . . .		186,279 mi/sec

Yes. Maxwell's equations implied that light was an electromagnetic phenomenon, a hitherto unsuspected form of the electric force.

The study of light was now to become a part of the study of electromagnetism.

But not everyone liked Maxwell's equations. Even Faraday was a bit piqued.

He wrote to Maxwell:

There is one thing I would be glad to ask you. When a mathematician engaged in investigating physical actions and results has arrived at his conclusions may they not be expressed in common language as fully, clearly, and definitely as in mathematical formulae? If so, would it not be a great boon to such as I to express them so? — translating them out of their hieroglyphics, that we also might work upon them by experiment. I think it must be so, because I have always found that you could convey to me a perfectly clear idea of your conclusions, which, though they may give me no full understanding of the steps of your process, give me results neither above nor below the truth, and so clear in character that I can think and work from them. If this be possible, would it not be a good thing if mathematicians, working on these subjects, were to give us the results in this popular, useful, working state, as well as in that which is their own and proper to them.

It wasn't until Helmholtz in 1871 decided to put all the competing theories in order that Maxwell's equations emerged as the prime candidate for the correct theory. Helmholtz's lab became the center for research into electromagnetic waves and the propagation of light.

Everyone agreed that light was a form of electric and magnetic interaction. . . . . but nobody could understand how it got from place to place!

The mechanism of the transmission of electric and magnetic forces was now a major problem. Everyone believed that some sort of medium (or substance) was necessary to support the fields.

" We have reason to believe, from the phenomenom of light and heat, that there is an aethereal medium filling space and permeating bodies. "

This was the famous luminiferous aether that was to occupy some physicists for the next 40 years.

Until Albert

Until Albert did away with it all.

The aether was supposed to fill all space...

... and had to have the contradictory properties:

1 completely permeable to material objects, while

2 at the same time, infinitely rigid in order to support the light properly.

But did the aether really exist?

In 1887 two U.S. Americans, A.A. Michelson and E.W. Morley, tried to detect the motion of the Earth through the aether using very sensitive apparatus.

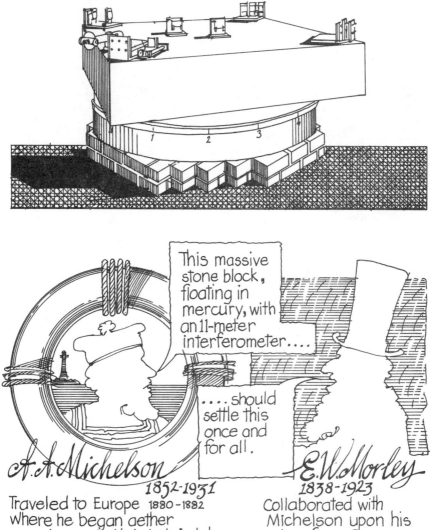

This massive stone block, floating in mercury, with an 11-meter interferometer....

.... should settle this once and for all.

A.A.Michelson
1852-1931
Traveled to Europe 1880-1882 where he began aether experiments in Helmholtz's lab.

E.W.Morley
1838-1923
Collaborated with Michelson upon his return from Germany.

They found **no** effect. The motion of the Earth through the aether was undetectable.

...already?

So, what did Albert do......

When Albert comes on the scene in 1895:

1 Hertz has experimentally verified Maxwell's equations

2 Marconi is busy trying to get money to build more wireless radios

3 The aether is assumed to exist but no one can find it.

Damn!

Albert does experiments to try to detect the aether.....

.....and nearly injures himself seriously.....

trying to push the apparatus beyond its limits.

•••• he wanted to understand what's going on when light propagates (spreads out) from place to place.

Like Faraday, Albert preferred simple pictures.

Remember, as a child Albert wondered how the compass needle could line up pointing to the North Pole without anything touching it.

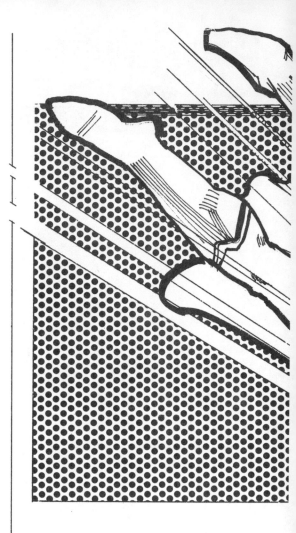

So Albert tried to form a simple picture of how light works.

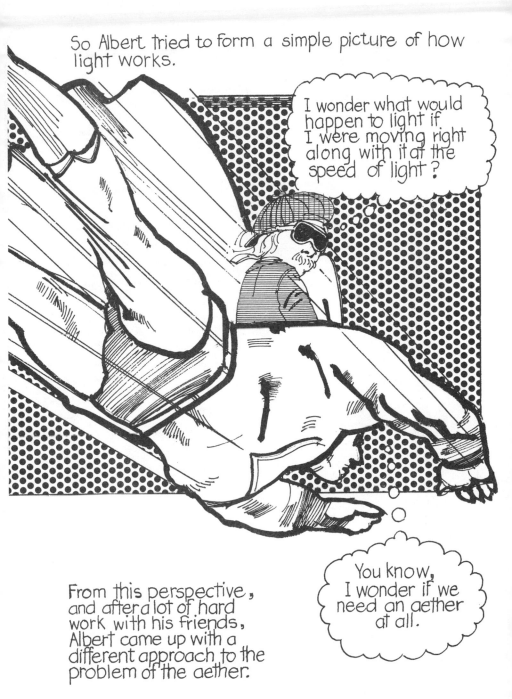

I wonder what would happen to light if I were moving right along with it at the speed of light?

You know, I wonder if we need an aether at all.

From this perspective, and after a lot of hard work with his friends, Albert came up with a different approach to the problem of the aether.

71

Of course we don't know **exactly** how it happened because altho'Albert could take an argument apart with just one punch, he didn't like to talk about it all that much.

Oh, the strong silent type, eh?

No. Albert never got used to being treated like a genius. He didn't like it. So he avoided going into detail about the way he thought about things.

And besides..... " In science.. ...the work of the individual is so bound up with that of his scientific contemporaries that it appears almost as an impersonal product of his generation. "

The key puzzler in his discussions with his friends was.... What exactly **would** happen if he rode along with a light wave at the speed of light?

Waves through the aether

Suppose I was holding a mirror ......

....and moving at the speed of light

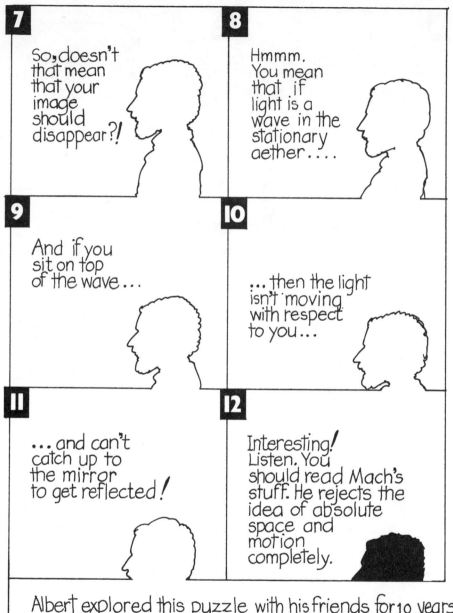

Albert explored this puzzle with his friends for 10 years, first at the ETH in Zurich from 1895-1900, and then at the Swiss Patent Office in Bern from 1901 to 1905.

When Albert graduated his ETH professors wouldn't recommend him, so he did odd teaching jobs for a year (he was a very good teacher) until Marcel Grossmann was able to pull some strings to get him a job at the Swiss Patent Office ...

..... a common civil service job for science graduates in those days.

I'm Conrad

I'm Maurice

And I'm Albert

In Bern he meets Maurice Solovine, and Conrad Habicht. They form the "Olympia Academy"...

... and along with Mileva Maric, Marcel Grossmann and Mike Besso, Albert continues to chew over that puzzle.

Moving with the speed of light, will my image disappear or not?

Hey! What about Mach?

Mach believed that a **physical** theory should be free of **metaphysical** constructions.

No one is competent to predicate things about absolute space and absolute motion; they are pure things of thought, pure mental constructs that cannot be produced in experience.

Ernst Mach 1838-1916

Mach also believed that a physical theory must be based only on primary sense perceptions (a belief that Lenin saw as creating political mischief later on). Albert benefited from Mach's willingness to challenge the accepted ideas of mechanics...

Mach's **Science of Mechanics** exercised a profound effect on me while I was a student.

I see Mach's greatness in his incorruptible skepticism.

Mach's ideas were useful because they helped Albert to reject the aether.

Since no one could find it anyway.

Here's what Albert thought...
No matter how it is that light gets from place to place (aether, shmaether) my image should **not** disappear.

But, then an observer on the ground would see the light leaving Albert's face at twice its normal velocity!

If I'm moving at 186,000 miles per second...

...and the light leaves my face at 186,000 miles per second...

...then relative to the ground the light should be moving at 186,000 + 186,000 = 372,000 miles per second! Right?

But that
didn't
make sense
either...

The speed of waves depended only on the medium and not on the source. For example, according to wave theory, sound from a passing train covers the distance to the observer in the same time no matter how fast the train is moving. And Maxwell's equations predicted the same thing for light. The observer on the ground should always see the light leaving Albert's face at the same speed no matter how fast Albert was moving.

But if the observer on the ground were to see the same speed for the light leaving Albert's face no matter how fast Albert were moving, then Albert should be able to catch up to the light leaving his face and his image should disappear.

But if his image shouldn't disappear, then light leaving his face should travel toward the mirror normally. But then the observer on the ground should see the light traveling toward the mirror at twice its normal speed. But if the observer on the ground . . . Oy veh!

Albert began to try to see if there
were any way for the speed
of light to be the **same** for **both**
the moving and the ground observers!

It nearly gave him a nervous breakdown...

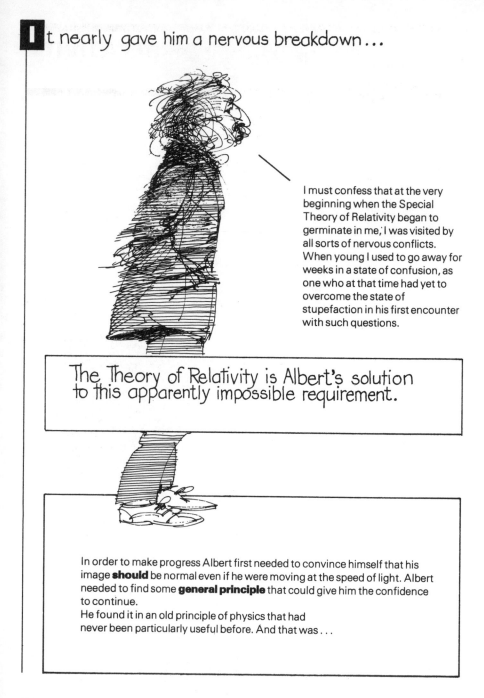

I must confess that at the very beginning when the Special Theory of Relativity began to germinate in me, I was visited by all sorts of nervous conflicts. When young I used to go away for weeks in a state of confusion, as one who at that time had yet to overcome the state of stupefaction in his first encounter with such questions.

The Theory of Relativity is Albert's solution to this apparently impossible requirement.

In order to make progress Albert first needed to convince himself that his image **should** be normal even if he were moving at the speed of light. Albert needed to find some **general principle** that could give him the confidence to continue.

He found it in an old principle of physics that had never been particularly useful before. And that was . . .

The
principle
of
relativity **?**

Galileo got into a lot of
trouble with the Inquistion.
His experiments on motion
led to the Principle of
Relativity.

Galileo 1564-1642

All steady
motion is
relative and
cannot be
detected
without
reference to
an outside
point.

83

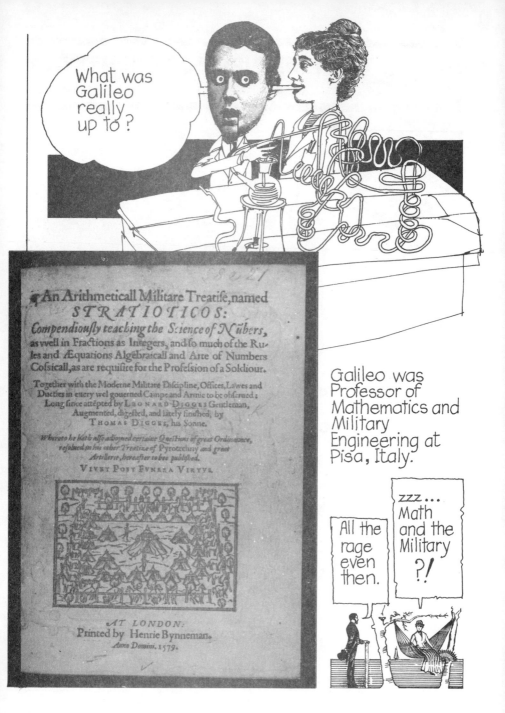

What was Galileo really up to?

An Arithmeticall Militare Treatise, named STRATIOTICOS: Compendiously teaching the Science of Nūbers, as well in Fractions as Integers, and so much of the Rules and Æquations Algebraicall and Arte of Numbers Cossicall, as are requisite for the Profession of a Souldiour.

Together with the Moderne Militare Discipline, Offices, Lawes and Dueties in every wel gouerned Campe and Armie to be obserued: Long since attēpted by LEONARD DIGGES Gentleman, Augmented, digested, and lately finished, by THOMAS DIGGES, his Sonne.

Whereto he hath also adioyned certaine Questions of great Ordinance, resolued in his other Treatise of Pyrotechny and great Artillerie, hereafter to bee published.

VIVIT POST FVNERA VIRTVS.

AT LONDON: Printed by Henrie Bynneman. Anno Domini. 1579.

Galileo was Professor of Mathematics and Military Engineering at Pisa, Italy.

All the rage even then.

zzz... Math and the Military ?!

84

Galileo worked on a lot of things. He built the first telescope in Italy and promptly sold it to the Doge of Venice for 1000 ducats and a life professorship.

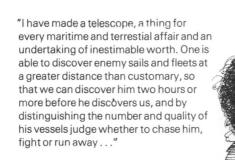

"I have made a telescope, a thing for every maritime and terrestial affair and an undertaking of inestimable worth. One is able to discover enemy sails and fleets at a greater distance than customary, so that we can discover him two hours or more before he discovers us, and by distinguishing the number and quality of his vessels judge whether to chase him, fight or run away . . ."

He also used the telescope to observe the moons of Jupiter. Being a practical man who needed money he tried to sell this first to the King of Spain and then to the States General of Holland as a navigational aid.

And, in addition, the discovery helped convince people that planets did revolve around the sun.

But Galileo's main concerns were with **terrestrial motion...**

Because of
    cannon balls.

Galileo took up from
Nicolo Tartaglia who had guessed
that the maximum range you
could get from a cannon
was to point it at 45°.

Galileo realized that the motion of projectiles could be analyzed by treating the horizontal and vertical motions separately.

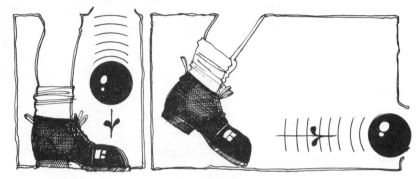

So if horizontal and vertical motion are combined this should mean that....

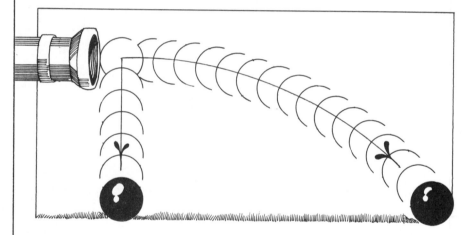

the cannonball fired from a perfectly horizontal cannon and another at the same time which falls vertically from the mouth of the cannon should hit the ground at the **same** time!

That's a strange result!

1 Doesn't the horizontal motion affect the vertical motion at all?

2 When I'm moving smoothly the cannonball's vertical motion **isn't** affected at all.

3 Galileo, then extended his argument to say that you couldn't use vertical motion or any other kind of motion to detect horizontal motion.

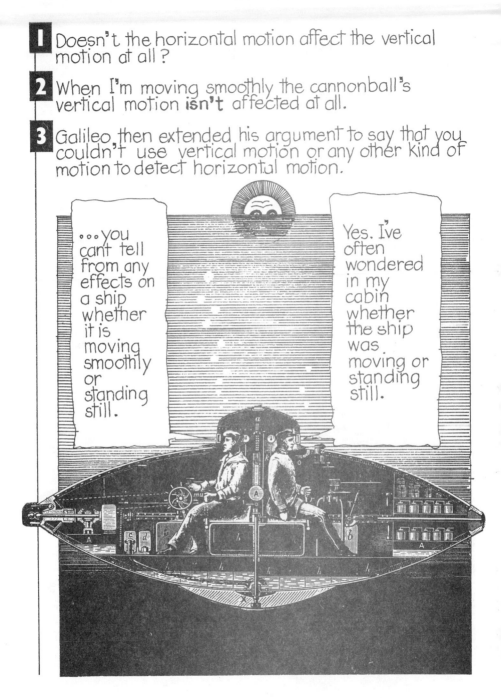

...you cant tell from any effects on a ship whether it is moving smoothly or standing still.

Yes. I've often wondered in my cabin whether the ship was moving or standing still.

And that's the principle of relativity. You can't tell if you're moving smoothly without looking outside.

The principle of relativity sounds harmless enough. Negating the idea of absolute rest wasn't a burning issue. But when applied to the problem of the aether it paved the way for the powerful arguments that became the Theory of Relativity

.... which first appeared in this magazine.

1905.   № 6.

ANNALEN
DER
PHYSIK.

F. A. C. GREN, L. W. GILBERT, J. C. POGGENDORFF, G. UND E. WIEDEMANN.

VIERTE FOLGE.

BAND 17.   HEFT 1.

KURATORIUM:
F. KOHLRAUSCH, M. PLANCK, G. QUINCKE,
W. C. RÖNTGEN, E. WARBURG.

UNTER MITWIRKUNG
DER DEUTSCHEN PHYSIKALISCHEN GESELLSCHAFT

M. PLANCK.

PAUL DRUDE.

LEIPZIG, 1905.
VERLAG VON JOHANN AMBROSIUS BARTH.

Based on the principle of relativity Albert argued he **should** be able to see his image normally even if he were moving at the speed of light.

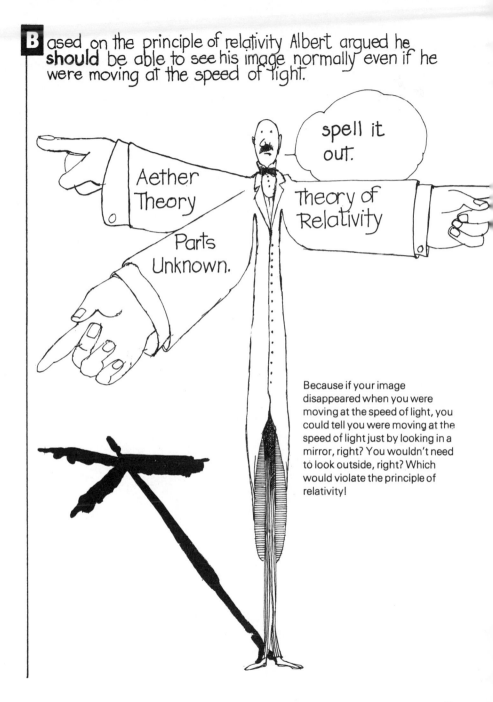

Aether Theory

Parts Unknown.

spell it out.

Theory of Relativity

Because if your image disappeared when you were moving at the speed of light, you could tell you were moving at the speed of light just by looking in a mirror, right? You wouldn't need to look outside, right? Which would violate the principle of relativity!

Impossible according to the principle of relativity.

Damn! There goes my image again. I keep telling them not to go 186,000 miles-per-second when I'm shaving.

That was half the problem solved. Albert's image **should** be normal. But could Albert see the light move away from his face at the speed of light relative to **him** . . . while, at the same time, observers on the ground would see the light leave Albert's face at the same speed of light relative to **them?**

How could this be possible?

Speed is distance divided by time (as in miles/hour). So Albert realized that if the **speed** were to be the same then the **distance** and **time** would have to be **different**. Which meant that there must be something suspect with time.

92

Perhaps the moving observer and the stationary observer observed **different times** ...

If both were to observe the **same** velocity for light.

I smell trouble here.

Because Albert took the principle of relativity as a starting point, he was led to rethink the concepts of space and time in order to make it come out all right.

This is how Albert finally expressed it in his Annalen der Physik article in 1905:

ON THE ELECTRODYNAMICS OF MOVING BODIES

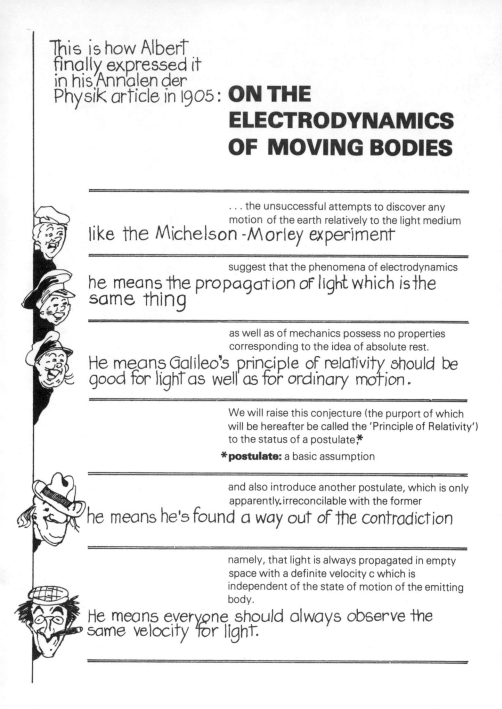

. . . the unsuccessful attempts to discover any motion of the earth relatively to the light medium

like the Michelson -Morley experiment

---

suggest that the phenomena of electrodynamics

he means the propagation of light which is the same thing

---

as well as of mechanics possess no properties corresponding to the idea of absolute rest.

He means Galileo's principle of relativity should be good for light as well as for ordinary motion.

We will raise this conjecture (the purport of which will be hereafter be called the 'Principle of Relativity') to the status of a postulate.*

*postulate: a basic assumption

---

and also introduce another postulate, which is only apparently irreconcilable with the former

he means he's found a way out of the contradiction

---

namely, that light is always propagated in empty space with a definite velocity c which is independent of the state of motion of the emitting body.

He means everyone should always observe the same velocity for light.

These two postulates suffice for the attainment of a simple and consistent theory of the electrodynamics of moving bodies based on Maxwell's theory for stationary bodies.

The introduction of a 'luminiferous aether' will prove to be superfluous inasmuch as the view here to be developed will not require an 'absolutely stationary space' provided with special properties . . .

He means he's doing away with the aether once and for all. Space will no longer require 'special properties' in order to transmit light.

But, certain conventional ideas about **time**, about **lengths**, about **mass**, about **velocity** had to be chucked out and replaced.

95

Albert's arguments are very simple because they are very logical. If you accept the two postulates Albert shows exactly how to make it come out O.K.

Albert was very pleased with the result. He wrote to his friend Conrad Habicht······

" *I promise you in return (if you write me) four works... The fourth work is based on concepts of the electrodynamics of moving bodies and modifies the theory of space and time, the purely kinematic part of this work should interest you.*"

SANFORD'S
MASS BLACK INK

Great ! He's really done it !

Now. Do you see what is happening? Albert says: no matter how light propagates when you are standing still....

Nice and sunny today. I think I'll go for a drive.

...it propagates exactly the same way when moving.

This is the principle of relativity, Albert's first postulate.

Such a nice day.

B ut Albert also says

" Light is always propagated in empty space with a definite velocity C which is independent of the state of motion of the emitting or receiving body."

An observer on the ground has to see light moving at the same velocity as the moving observer. This is Albert's 2nd postulate.

Remember the compass?

Albert wondered how the compass needle interacted with the Earth's magnetism.

How do magnetic (or electric) effects get transmitted from one place to another?

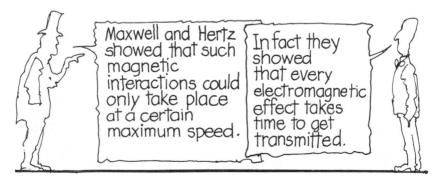

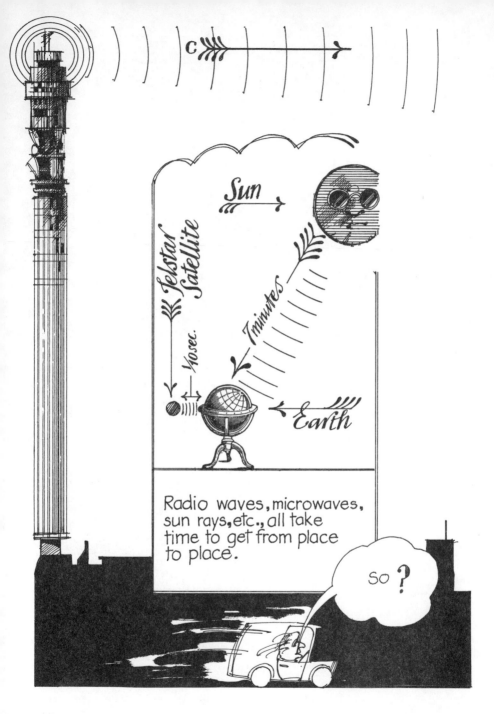

Radio waves, microwaves, sun rays, etc., all take time to get from place to place.

So Albert made an inference. Based on the experience with electricity as summarized by Maxwell and verified by Hertz, Albert proposed that there are no instantaneous interactions **at all** in nature.

Here is the simple **physical** meaning of Albert's 2nd postulate:

Every interaction takes time to get from one place to the next.

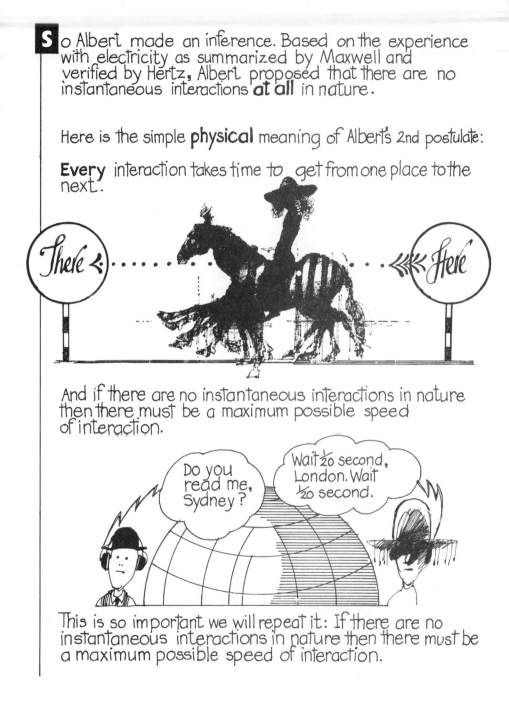

And if there are no instantaneous interactions in nature then there must be a maximum possible speed of interaction.

This is so important we will repeat it: If there are no instantaneous interactions in nature then there must be a maximum possible speed of interaction.

The maximum possible speed of interaction in nature is the speed of the electromagnetic interaction – which is the speed of light!

It's quite revolutionary really.

Now by the principle of relativity, the maximum speed of interaction must be the same for every observer no matter how they are moving.

otherwise you could tell you were moving simply by measuring the speed of light.

The speed of light (the maximum speed of interaction) is a universal constant. This is Albert's 2nd postulate.

Everyone sees the same speed for light no matter
how they are moving.

104

The maximum possible speed is a material property of our world.

But how is it possible ?

Well... Albert has to show that something unexpected is going on.

Albert has to show:

1 How everyone can see the same speed for light (c).

and

2 What happens when you try to get an object to move faster than c.

To do this Albert shows that:

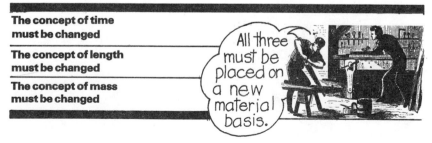

The concept of time must be changed

The concept of length must be changed

The concept of mass must be changed

All three must be placed on a new material basis.

105

So this is Albert's position:

1 There are no instantaneous interactions in nature.

2 Therefore there must be a maximum possible speed of interaction.

3 The maximum possible speed of interaction is the speed of the electromagnetic interaction.

4 The speed of the electromagnetic interaction is the speed of light.

5 The speed of light is the maximum possible speed.

The really difficult part was showing how everyone could see the same speed for light.

Let's see how he did it.

Albert nearly drove himself crazy until he realized that TIME was the joker in the pack! The time elapsed between events was not necessarily the same for all observers!

Remember speed is distance gone divided by the time it takes.
In symbols: $S = \frac{D}{T}$

So the moving person could observe the light traveling a certain distance D in a certain time T to give the speed of light c ....

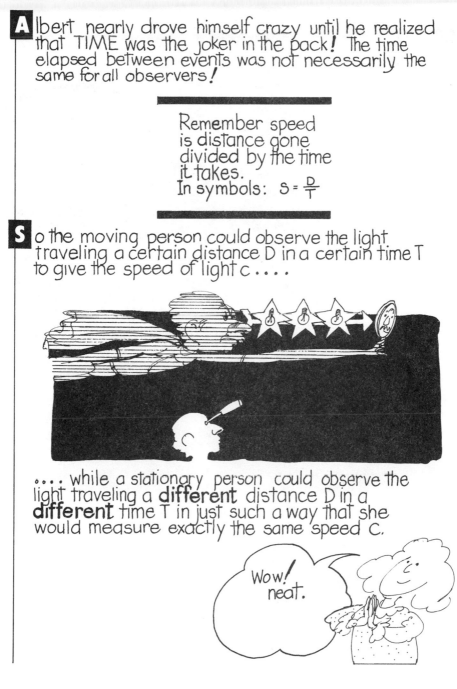

.... while a stationary person could observe the light traveling a **different** distance D in a **different** time T in just such a way that she would measure exactly the same speed C.

Wow! neat.

It is neat. Here's how Albert analyzed the phenomenon of simultaneous events.

Simultaneous . . . .  . . . . events?

Yes. Albert points out that any measurement of time uses the idea of simultaneous events.

We have to understand that all our judgments in which time plays a part are always judgments of simultaneous events. If, for instance, I say "That train arrives here at 7 o'clock" I mean something like this: "The pointing of a small hand of my watch to 7 and the arrival of the train are simultaneous events."

So?

Albert argued that simultaneous events in one frame of reference would not necessarily be simultaneous when viewed from a different frame.

Albert called this the **RELATIVITY OF SIMULTANEITY**

Albert suggests that we try to picture his argument in terms of a train....

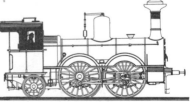

.... as the moving frame of reference and the railway embankment as the stationary frame of reference.

Now we can put them together. Let's have a passenger car too, Mike.

There. Now imagine that someone in the center of the passenger car holds a device which can send out a beam of light in the forward direction and at the same time a beam of light in the backward direction.

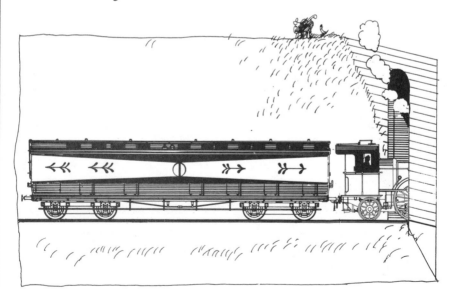

And we further imagine that the front door and back door can be opened automatically by the light beams.

Then to the person holding the device the doors of the passenger car will open simultaneously. But to a person on the embankment, Albert argues, the back door will open before the front door!

See? Because for the stationary person, the back door moves forward to meet the light pulse, while the front door moves away from the light pulse.

But which is it? Do the doors open at the same time or don't they?

That's the point. Since the speed of light is to be the same for both frames, Albert argues that. . . .

Events which are simultaneous with reference to the train are not simultaneous with respect to the embankment and vice versa.

You better give us a chance to get used to this.

Take a more common sense example: distance traveled.

Imagine that our person in the middle of the carriage gets up and goes to the front door.

Hang on.

OKay.

V⟶

Now, how far has our imaginary person gone?

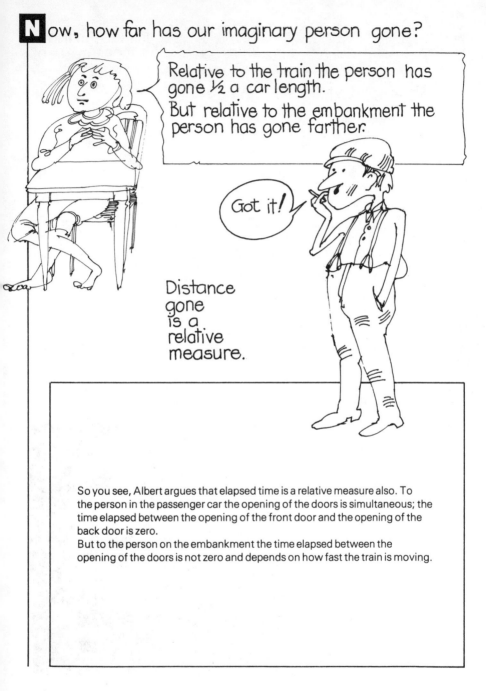

Relative to the train the person has gone ½ a car length.
But relative to the embankment the person has gone farther.

Got it!

Distance gone is a relative measure.

So you see, Albert argues that elapsed time is a relative measure also. To the person in the passenger car the opening of the doors is simultaneous; the time elapsed between the opening of the front door and the opening of the back door is zero.
But to the person on the embankment the time elapsed between the opening of the doors is not zero and depends on how fast the train is moving.

Alright. What's next?

Next, Albert argues, is the relativity of the measurement of length.

Albert asks, what is the length of the passenger car?

An observer in the train measures the interval by marking off his measuring rod in a straight line.

(This is the length measured by the **moving** observer.)

But it is a different matter when the distance has to be judged from the embankment.

Right. Albert argues that to measure the length of the car as seen from the embankment, we have to mark the positions on the embankment which are being passed by the front door and the back door at the same time T - **as judged from the embankment**. The distance between these points is **then** measured with a measuring rod.

(This is the length of the car as measured by the **stationary** observer)

A lbert says:

It is by no means evident that this last measurement will supply us with the same result as the first.

Thus, the length of the train as measured from the embankment may be different from that obtained by measuring in the train itself.

Albert is preparing the ground for a reconsider-
ation of Newton's analysis of space time & motion.

Classical mechanics assumes that: **1** The time interval between events is independent of the motion of the observer.

2 The space interval (length) of a body is independent of the motion of the observer.

Unjustifiable!

Space and time intervals are relative and do depend on the motion of the observer.

Newton says:
Space and time intervals are absolute and the speed of light is relative.

Albert says:
The speed of light is absolute and space and time intervals are relative.

Space and time relative? Absolute nonsense.

Albert replaces Newton's metaphysical absolutes, the constructs of absolute space and time, with a **material** absolute: there are no instantaneous interactions in nature!

Albert's contribution was dramatic because it so fundamentally challenged the framework of classical physics that had been accepted for the previous 200 years.

classical = perfect

So?
How does this affect us?

Quite right. There's no need to get **that** excited about relativity just because a bunch of physicists got excited by it.

Relativity theory had nothing to do with the development of the A-bomb. *The Anti-Nuclear Handbook* tells the story. And we'll discuss this again later.

Meanwhile let's see what the rest of Albert's argument consists of.

Albert didn't just argue that space and time intervals needed to be reformulated. He showed exactly how to do it.

Albert's program:

1 To find a place and time of an event relative to the railway embankment when we know the place and time of the event with respect to the train

It's impossible

such that

2 Every ray of light possesses the speed c relative to **both** the embankment and the train.

"This question leads to a quite definite positive answer."

Since we are talking about measurements of distances and time, we are talking about numbers. Albert needs to use the traditional language of numbers to make it come out right.

So we'd better have a look at a bit of mathematics first to see how that all got started and how it relates to physics

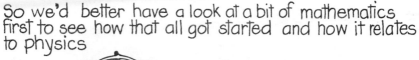

The first step of course was counting.

There are at least 7 more dinosaurs around here. We'd better tell the others.

Tallying has been dated to 30,000 B.C. They used scratches on bones to do it.

And the next big step was measurement, which got its real start with the rise of the cities.

The Egyptian ruler-priests needed measures of distance, area, volume and weight to assess taxes and run the state.

To keep records of what they were doing they had to write down the accounts. So written numerals were the next step. And this is where mathematics began to get mystified. Because the priests kept writing for themselves.

Hiero-glyph = priest's writing

Anyway, the Babylonian and Sumerian priests got rather good at arithmetic starting about 3000 B.C.

At first they wrote their numbers like this

$\vee$ = 1   and $\langle$ = 10

so a number like 59 would be written

$\langle\langle\langle$ ... = 59

But later the Babylonians developed the first place system for numbers.

They used a base of 60

$\Upsilon\Upsilon$ $\Upsilon\Upsilon$ $\Upsilon\Upsilon$

2×60×60 + 2×60 + 2 = 7322

or : 7322 = 7×(10×10×10) + 3×(10×10) + 2×10 + 2

The Babylonians had as good a computation system as ours.

You owe me
$\Upsilon\Upsilon$ $\Upsilon\Upsilon$
bushels
of wheat.

Thats 122
bushels
too many.

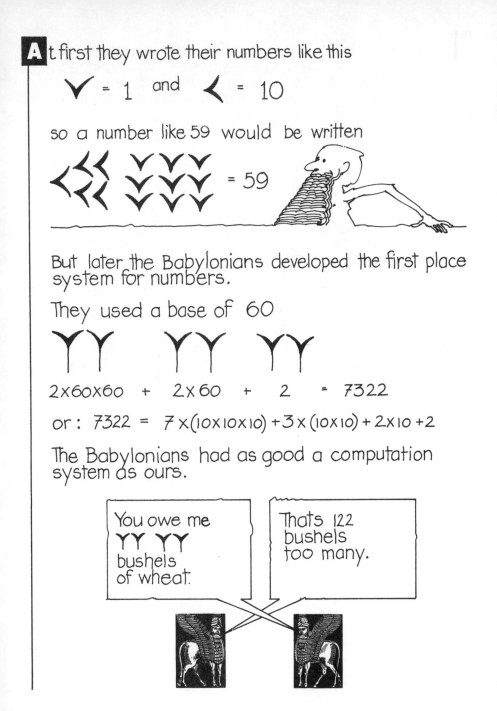

Now a skill developed in isolation for centuries by a special group of people may become somewhat boring. By 1900 B.C. the Babylonians had made up lots of little problems for their own instruction and amusement.

This was the beginning of **ALGEBRA**

They wrote it all down on clay tablets

Babylonian tablet 1500 B.C. with algebra equation on it.

Of course it was not exactly what we now use. The Babylonians didn't have algebraic notation. (That had to wait for the rise of the Islam and Hindu merchant class.)

What the Babylonians did was to pose an abstract problem....

Find the side of a square if the area less than the side is 14×60+30

....and then give the detailed steps to the solution

Take half of one and multiply by half of one.

Add this to 14×60+30. This is the square of $29 + \frac{30}{60}$.

Now add half of one and the result is 30, the side of the square.

While what we do now is write

$$x^2 - x = 870 \quad \rightarrow \quad x = \frac{1}{2} + \sqrt{(\frac{1}{2})^2 + 870} = 30$$

There's not much difference really. In fact we solve equations on modern computers with exactly the same step-by-step method first used by the Babylonian priests.

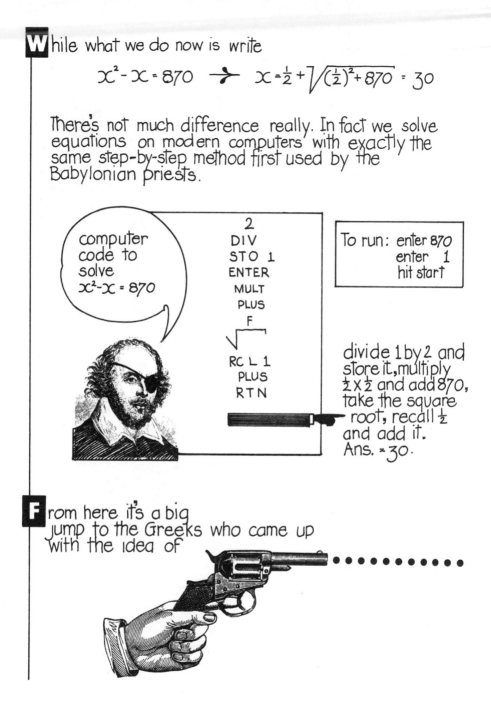

computer code to solve $x^2 - x = 870$

```
       2
     DIV
     STO 1
   ENTER
    MULT
    PLUS
      F
   √⌐
 RCL 1
   PLUS
    RTN
```

To run: enter 870
enter 1
hit start

divide 1 by 2 and store it, multiply ½ x ½ and add 870, take the square root, recall ½ and add it.
Ans. = 30.

From here it's a big jump to the Greeks who came up with the idea of

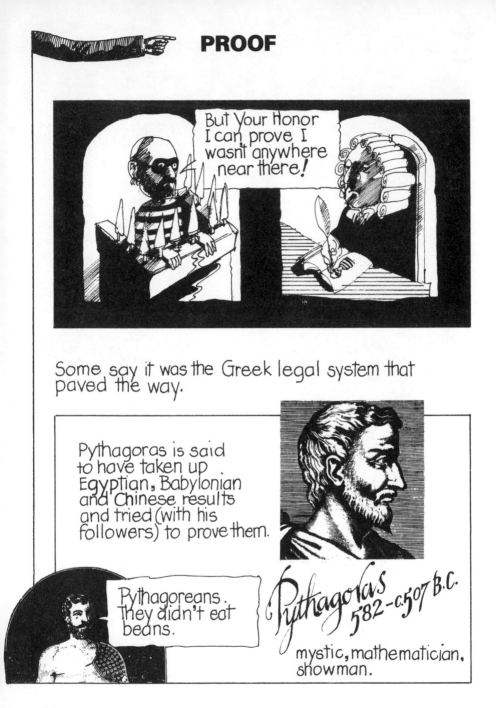

But Your Honor I can't prove I wasn't anywhere near there!

Some say it was the Greek legal system that paved the way.

Pythagoras is said to have taken up Egyptian, Babylonian and Chinese results and tried (with his followers) to prove them.

Pythagoreans. They didn't eat beans.

Pythagoras
582 – c.507 B.C.

mystic, mathematician, showman.

A famous example is the Pythagorean Theorem. Remember this from school?

C square on the hypotenuse

square on side **B**

square on side **A**

The square on the hypotenuse equals the sum of the squares on the other two sides.

We write it nowadays as $C^2 = A^2 + B^2$.

... and we mean: Take the length of side C and multiply it by itself. This gives the area of square C. Do the same for squares A and B.

Then $C^2 = A^2 + B^2$ which is saying it algebraically.

Keep this in mind – Albert will use it later.

Greek mathematicians labored for centuries trying to trisect an angle with only a compass and a straightedge....

Why don't we just measure it?

You're missing the point!

.... and that was where matters stood untill the Hindus invented our modern algebra.

Aryabhata (A.D. 470) wrote down all the Hindu methods of multiplication, long division and algebra that we use today. They made up exercises (like the Babylonians) to help them with calculations of taxation, debt and interest.

A merchant pays duty on certain goods at 3 different places.

At the first he gives $\frac{1}{3}$ of his goods, at the second $\frac{1}{4}$ of what he has left and at the third $\frac{1}{3}$ of the remainder. The total equals 24 coins.

What had he at first?

x = what he had at first

1   gives up $\frac{1}{3}x$

2   gives up $\frac{1}{4}(\frac{2}{3}x)$

3   gives up $\frac{1}{3}(\frac{3}{4}(\frac{2}{3}x))$

24 = $\frac{1}{3}x + \frac{1}{6}x + \frac{1}{6}x$

x = 36 coins

Meanwhile Medieval Europe wallowed in the throes of the Age of Faith until ....

POW! The Renaissance.

Rebirth to you

Mary is 24 years old. Mary is twice as old as Ann was when Mary was as old as Ann is now. How old is Ann?

ask Ann!

Now improved mathematics was needed for astronomy, for navigation, for gunnery, for shipbuilding, for hydraulic engineering, for building technology.

So there came:

disputed priority

Algebraic notation	Vieta (1580)
Decimals	Stevinus (1585)
Logarithms	Napier (1614)
Slide rule	Gunter (1620)
Analytic geometry	Descartes (1637)
Adding machine	Pascal (1642)
Calculus	Newton (1665)
Calculus	Leibniz (1684)

233

Of course there has been a long history of number mystics who were very impressed with their own cleverness . . . .

Ann is 18 years old

Pythagoras: "Bless us divine number, who generated Gods and men. Number containest the root and source of eternally flowing creation."

Plato: "God ever geometrizes"

Galileo: "The book of the Universe is written in mathematical language, without which one wanders in vain through a dark labyrinth."

Hertz: "One cannot escape the feeling that these mathematical formulas have an independent existence and intelligence of their own, that they are wiser than we are, wiser even than their discoverers, that we get more out of them than was originally put into them."

. . . . and who forgot the original impulses that led them to mathematics in the first place.

134

But in reality mathematics is only a **language** invented by human beings to describe sizes and quantities and relationships between measurable things.

And that's exactly how Albert used math .... to express the relationship between the place and time of an event in relation to the embankment when we know the place and time of the event with respect to the train.

And now let's have that passenger car again Mike.

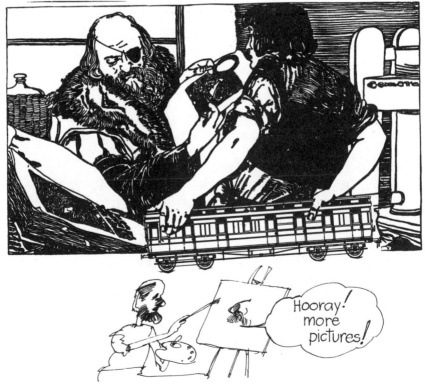

Hooray! more pictures!

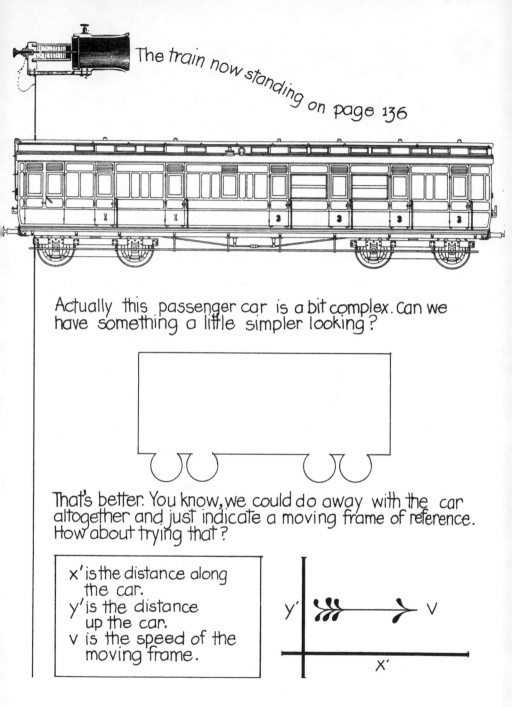

Actually this passenger car is a bit complex. Can we have something a little simpler looking?

That's better. You know, we could do away with the car altogether and just indicate a moving frame of reference. How about trying that?

x′ is the distance along the car.
y′ is the distance up the car.
v is the speed of the moving frame.

y′   → v

x′

There, that's simpler. Now we have a moving frame of reference y'x'.

And a stationary frame of reference x y.

---

x is the distance along the embankment

y is the distance up the embankment.

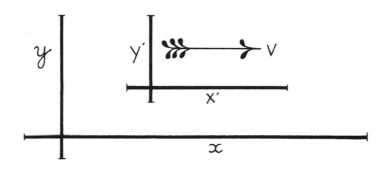

Which corresponds to the passenger car and the embankment. We mark an event in the moving frame by its coordinates y'x' and time t' and we mark the same event in the stationary frame by its coordinates y x and its time t.

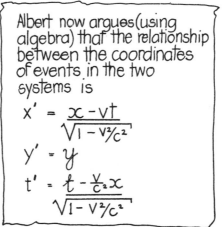

Albert now argues (using algebra) that the relationship between the coordinates of events in the two systems is

$$x' = \frac{x - vt}{\sqrt{1 - v^2/c^2}}$$

$$y' = y$$

$$t' = \frac{t - \frac{v}{c^2}x}{\sqrt{1 - v^2/c^2}}$$

137

The system of equations on page 137 is known by my name.

H. A. Lorentz 1853-1928
Dutch theoretical physicist
discovered the
Lorentz transformation,
senior statesman of
physics and friend
of Einstein.

Right. Now we must show what's going on here • • • • •

Imagine that both frames of reference are at rest (relative to each other of course).

And we have two identical rather special light clocks in them (designed by the U.S. physicist R.P. Feynman).

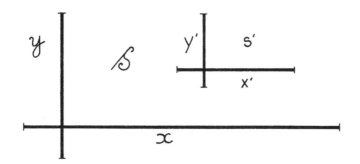

I smell something fishy here.

The light bulb gives out regular pulses of light which go up to the mirror, get reflected and bounce back to the counter which goes click! click!

Now we imagine that the s' system is given a velocity v so that it is a moving system with respect to the system s.

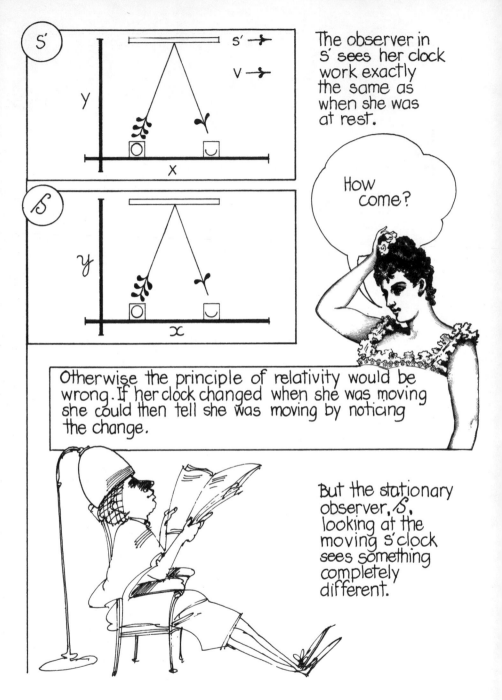

The observer in S' sees her clock work exactly the same as when she was at rest.

How come?

Otherwise the principle of relativity would be wrong. If her clock changed when she was moving she could then tell she was moving by noticing the change.

But the stationary observer, S, looking at the moving S' clock sees something completely different.

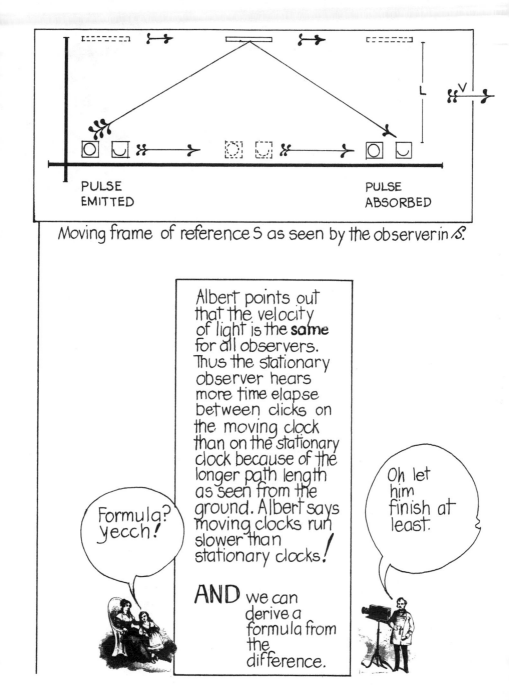

PULSE
EMITTED

PULSE
ABSORBED

Moving frame of reference S as seen by the observer in S̄.

Albert points out that the velocity of light is the **same** for all observers. Thus the stationary observer hears more time elapse between clicks on the moving clock than on the stationary clock because of the longer path length as seen from the ground. Albert says moving clocks run slower than stationary clocks!

AND we can derive a formula from the difference.

Formula? yecch!

Oh let him finish at least.

141

Don't have a nervous breakdown.

a **go slowly**

b **use pencil and paper**

c **get a friend to come along!**

The Key Terms:
<hr/>

v = the speed of the moving frame
t' = the time between clicks in the moving frame
t = the time between clicks in the stationary frame
c = the speed of light

1

The time, t', between clicks in the moving frame is the time the light takes to reach the mirror L/c plus the time it takes to return, again L/c.

$$\text{so } t' = \frac{2L}{c}$$

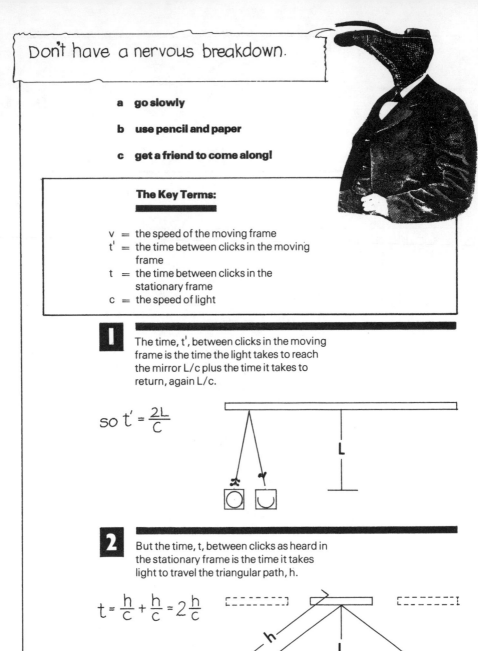

2

But the time, t, between clicks as heard in the stationary frame is the time it takes light to travel the triangular path, h.

$$t = \frac{h}{c} + \frac{h}{c} = 2\frac{h}{c}$$

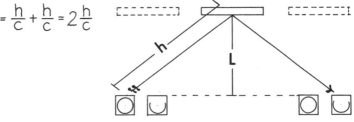

3 Now in the time t, the moving frame moves a distance d. And d = vt.

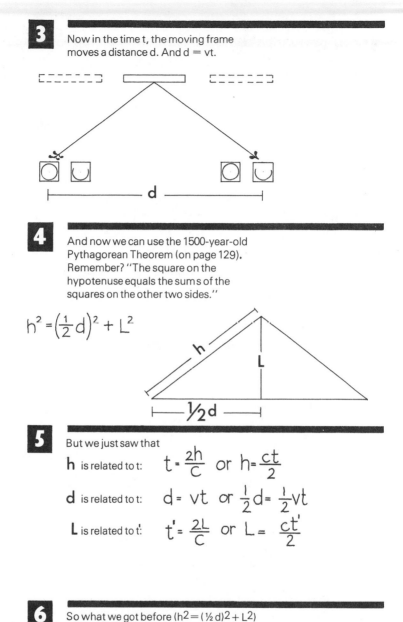

4 And now we can use the 1500-year-old Pythagorean Theorem (on page 129). Remember? "The square on the hypotenuse equals the sums of the squares on the other two sides."

$$h^2 = \left(\tfrac{1}{2}d\right)^2 + L^2$$

5 But we just saw that

h is related to t: $\quad t = \dfrac{2h}{c} \quad$ or $\quad h = \dfrac{ct}{2}$

d is related to t: $\quad d = vt \quad$ or $\quad \tfrac{1}{2}d = \tfrac{1}{2}vt$

L is related to t': $\quad t' = \dfrac{2L}{c} \quad$ or $\quad L = \dfrac{ct'}{2}$

6 So what we got before $(h^2 = (\tfrac{1}{2}d)^2 + L^2)$ can now be substituted for:

$$\left(\frac{ct}{2}\right)^2 = \left(\frac{1}{2}vt\right)^2 + \left(\frac{ct'}{2}\right)^2$$

143

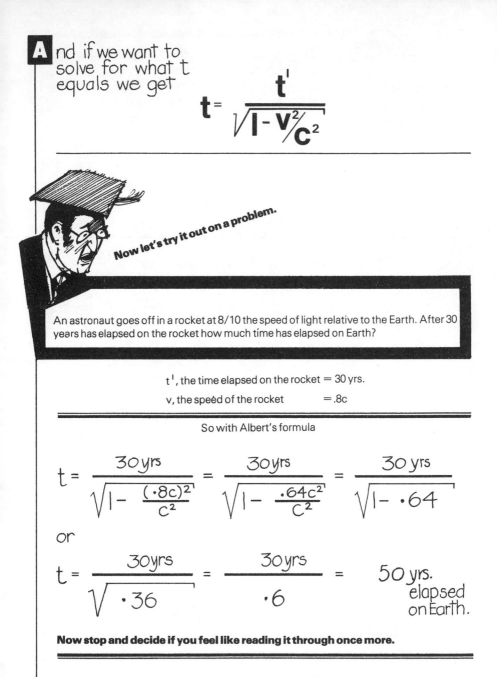

A nd if we want to solve for what t equals we get

$$t = \frac{t^1}{\sqrt{1 - v^2/c^2}}$$

Now let's try it out on a problem.

An astronaut goes off in a rocket at 8/10 the speed of light relative to the Earth. After 30 years has elapsed on the rocket how much time has elapsed on Earth?

t^1, the time elapsed on the rocket = 30 yrs.

v, the speed of the rocket = .8c

So with Albert's formula

$$t = \frac{30\,yrs}{\sqrt{1 - \frac{(\cdot8c)^2}{c^2}}} = \frac{30\,yrs}{\sqrt{1 - \frac{\cdot64c^2}{c^2}}} = \frac{30\,yrs}{\sqrt{1 - \cdot64}}$$

or

$$t = \frac{30\,yrs}{\sqrt{\cdot36}} = \frac{30\,yrs}{\cdot6} = 50\,yrs.\ \text{elapsed on Earth.}$$

Now stop and decide if you feel like reading it through once more.

Albert's conclusion is...

"As judged from s the clock is moving with velocity v; so the time which elapses between two strokes of the clock is not one second but $\dfrac{1}{\sqrt{1-v^2/c^2}}$ seconds.

i.e. a somewhat longer time. As a consequence of the motion, the clock goes more slowly than when at rest."

Whew, this is ridiculous.

Well, you did want to find out about relativity, didn't you?

What Albert achieved was a glimpse into how the world looks when things move at close to the speed of light. This is so far removed from everyday experience that it takes a certain amount of work to visualize it.

You can say that again!

But remember, Albert was led to this picture by a desire to understand how electric and magnetic forces propagate. He realized that the new area of experience represented by Maxwell's equations required deep modifications of the ideas based on the old area of experience represented by Newton's laws.

Now all we have to show is how the velocities come out right.

Velocities?

Yes, you remember every observer must see the same velocity of light no matter how they are moving (in a steady way of course).

Mike, lets have our passenger car again.

Good. Now we imagine that our person in the middle of the car gets up and walks to the front door at a rate of w = 3 miles per hour. We further imagine that the train is moving at a velocity of v = 20 miles per hour.

so?

Well, how fast is
our person moving U. with
respect to the embankment?

$$U = V + W = 20 + 3 \quad \text{m.p.h.} \; ?$$

That's right (almost).
But Albert tells us that
the distances and times
measured on the train
are not the same as the
distances and times
measured on the
embankment.

So
what
do
we
do?

Well, to take relativity into account
we just have to be very precise.
In reality when we say that a
person walks at 3 miles per hour
with respect to the train we mean
that they cover the distance
to the front door X in a time t where
x and t are measured on the train, right?

And we know that distances and times as measured on the train are not the same as when measured from the embankment, right?

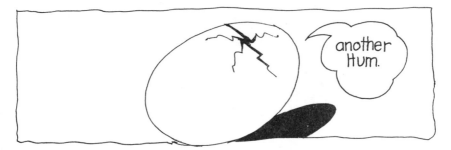

Hum.

So what we need to do is to convert x' and t' as measured on the train into x and t as measured on the embankment.

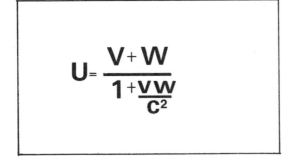

another Hum.

Doing this Albert shows that the velocity U of the person as seen from the ground is given by

$$U = \frac{V + W}{1 + \dfrac{VW}{C^2}}$$

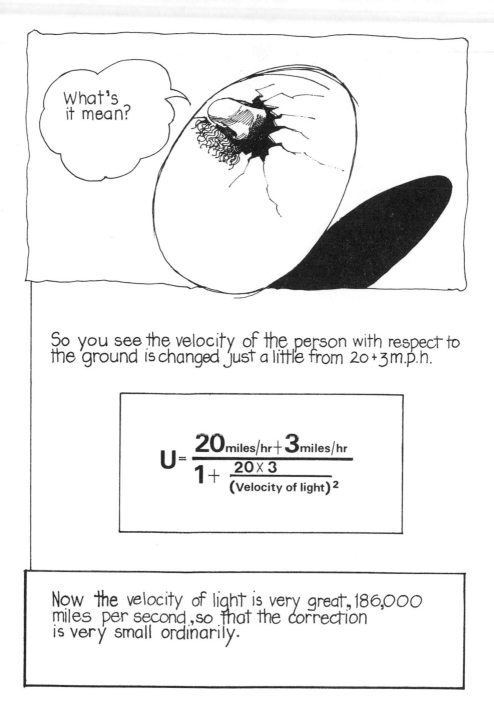

So you see the velocity of the person with respect to the ground is changed just a little from 20+3 m.p.h.

$$U = \dfrac{20_{\text{miles/hr}} + 3_{\text{miles/hr}}}{1 + \dfrac{20 \times 3}{(\text{Velocity of light})^2}}$$

Now the velocity of light is very great, 186,000 miles per second, so that the correction is very small ordinarily.

But let's try the formula when the train goes at the speed of light.

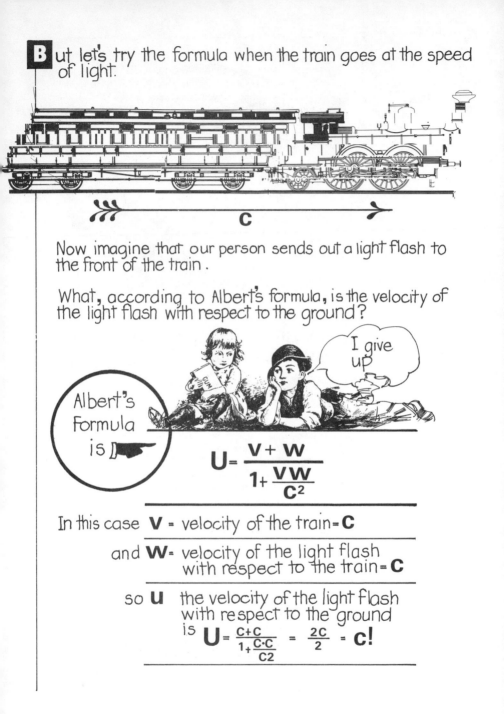

Now imagine that our person sends out a light flash to the front of the train.

What, according to Albert's formula, is the velocity of the light flash with respect to the ground?

I give up

Albert's Formula is

$$U = \dfrac{V + W}{1 + \dfrac{VW}{C^2}}$$

In this case **V** = velocity of the train = **C**

and **W** = velocity of the light flash with respect to the train = **C**

so **U** the velocity of the light flash with respect to the ground is

$$U = \dfrac{C + C}{1 + \dfrac{C \cdot C}{C2}} = \dfrac{2C}{2} = C!$$

It's a neat formula. Albert has shown that his proposed modifications of space and time intervals lead to a new formula for the addition of velocities. The new formula expresses the new fact: there are no instantaneous interactions in nature, nothing can go faster than the speed of light.

Magic!

It takes a lot of work to follow this.

Don't get worried. Among physicists there's a saying: "You never really understand a new theory. You just get used to it."

Understanding is based on experience and it is difficult to accumulate experience about things moving near the speed of light! (Unless you're a worker on high-speed particles.)

Albert now has to show what happens when you try to get an object to exceed the speed of light.

This is how Albert argues:

To get an object moving you've got to apply a **Force**.

Force   strength, power 13c.

force?   body of
            armed men   14c.

            strong, producing
            a powerful effect 16c.

            From Latin
            A.D. 200 or earlier
            fortis strong

push or a pull

or a kick

or a **hit**

In physics force is another word for•••

•••interaction!

To get an object moving really fast you've got to give it lots of "hits"

or a constant steady push, say, by an engine. There are lots of practical difficulties in applying a large steady force to an object. Air resistance. Mechanical breakdown. Running out of fuel.

But Albert is concerned with a deeper difficulty. If there are no instantaneous interactions in nature and if the speed of light is the fastest you can go, what exactly does happen when an object starts to appoach the speed of light?

Wow! Does it explode?!?

No. Wait and see.

We imagine we apply a steady force to a particle (which we call an electron).

No. No. Electrons are much smaller. Oh well. Never mind.

When an object picks up speed we say it accelerates.

It was Newton who postulated a connection between force and acceleration.

Hey, what about Mach's and Hertz's criticism?

Oh, stop showing off!

Newton said F=ma. Or a=F/m. The acceleration, a, is proportional to the applied force, F, and is inversely proportional to the mass, m (also called the inertia) of the object.

The bigger the force the faster it picks up speed. The bigger the mass or inertia the harder it is to get it moving fast.

Some call it the "power to weight ratio."

It is easier to get a light car rolling than a loaded truck.

But we'll return to the concept of mass or inertia in a moment.

...leading to $E = mc^2$?

Yep! But let's analyze the motion of the electron first!

1 If the electron is at rest then its subsequent motion is given by $F = ma$.

2 But suppose the electron already has a speed v?

Then the electron is at rest with respect to a frame of reference S' moving with velocity v with respect to S.

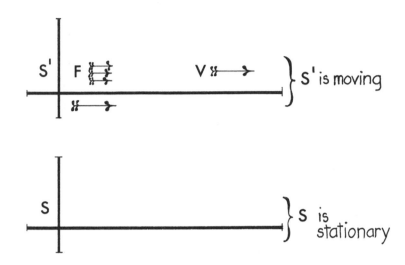

Relative to S', the electron has an acceleration $a = F/m$ (because the electron is at rest relative to S').

Right. Albert knows how to find the place
and time of an event with respect to the
embankment S, when he knows the place
and time of the event with respect to the train S'.

The event in this case is the acceleration of the electron.

Here's what happens:

---

1 The electron goes faster because of the force

---

but

---

2 In the frame where the electron is at rest the *time* over which the force acts gets smaller and smaller compared to the stationary frame (moving clocks run slow, remember?)

---

so

---

3 In the frame where the electron is at rest the force acts for a shorter and shorter time, the closer the electron gets to the speed of light. As seen from the ground the electron hardly has time to get pushed at all!

Wow! You give relativity your little finger and it takes your whole arm!

Albert expresses the process by a concrete new formula.

EINSTEIN'S FORMULA 1905

$$a = \frac{F}{M}\left(1 - \frac{V^2}{C^2}\right)^{3/2}$$

My new one (1905)

Compared to my old one!

NEWTON'S FORMULA 1686

$$a = \frac{F}{M}$$

Once again,
the new formula
re-expresses the new fact:
There are no instantaneous interactions in nature.
Nothing can go faster than the speed of light.

**Albert's formula shows that when
v=c, a=zero! So even if you
keep on pushing, the electron ·
doesn't pick up any more speed.**

The meaning is **'relatively'** straightforward.

1 If you push on an object with a force and it hardly picks up any speed at all, you say it has a lot of inertia!

2 Thus as the electron approaches the speed of light it appears to get heavier and heavier because it becomes harder and harder to increase its speed.

lbert now wants to show how the speed of the electron is related to its energy.

Ah, energy!

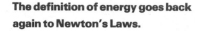

The definition of energy goes back again to Newton's Laws.

1 When a force, F, acts on a body of mass, m, for a distance, d, it is useful to say that work, W, has been done on the body.

2 The work, W, is *assigned* a value $W = Fd$.

3 By using $F = ma$ you can show that the work as *defined* by $W = Fd$ is exactly equal to $\frac{1}{2}mv^2$.

4 The expression $\frac{1}{2}mv^2$ is also given a name. It is called the **kinetic energy** of the body.

6 The more work (Fd) you put into pushing a body, the more kinetic energy ($\frac{1}{2}mv^2$) it gets.

It's all a naming game connected up by $F = ma$!

Albert now says "Wait a minute." We can put in the work ($W = Fd$) but the body doesn't pick up speed in the same way. Why? Because now

$$F = \frac{ma}{\left(1 - \frac{v^2}{c^2}\right)^{3/2}}$$

So Albert's modification leads to a new formula. The work now equals:

$$W = \frac{MC^2}{\left(1 - \frac{v^2}{C^2}\right)^{1/2}} - MC^2 \quad \text{vs} \quad W = \frac{1}{2}mv^2$$

Albert's formula　　　　　　Newton's formula

Albert is satisfied. He concludes . . . .

When v = c, W becomes infinite. Velocities greater than that of light have — as in our previous results — no possibility of existence.

Albert argues that as you give an object more and more energy . . . . .

faster faster

. . . . instead of going faster and faster it gets heavier and heavier!

So even if you gave a rocket 1,000,000,000,000,000,000,000,000,000,000,000,000,000,000,000,000,000,000,000,000,000,000,000,000,000,000,000 foot pounds of thrust, it would still be going less than the speed of light!

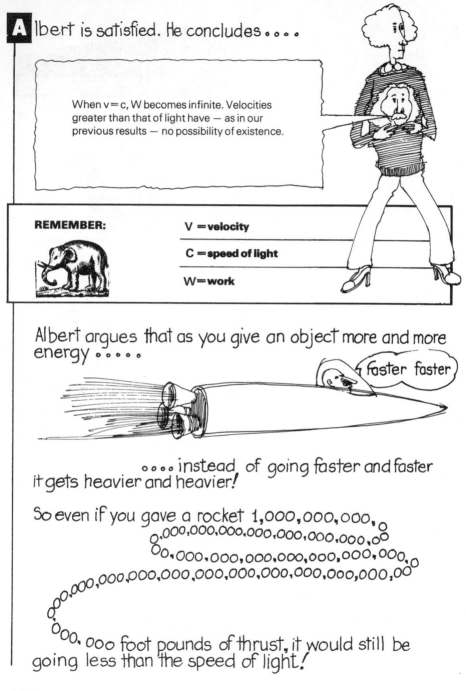

But that's not all.
If work goes into
giving the
body more inertia...

...then inertia
must contain **energy!**

What?
dead weight
has
energy?!?

Yes. Albert says we need a new definition of energy. The old Newtonian one
(k.e. $= \frac{1}{2} mv^2$) is only good for speeds much less than the speed of light.

So...

1	Albert has shown (page 161) that the work W equals $\dfrac{mc^2}{(1 - V^2/c^2)^{\frac{1}{2}}} - mc^2$
2	So Albert says let's call the quantity $\dfrac{mc^2}{(1 - V^2/c^2)^{\frac{1}{2}}}$ the energy E of the electron.
3	Then, with this definition of energy, Albert's formula reads $E = W + MC^2$

What Albert says is...even if W= zero. If you
don't put in any work **at all,**
then the electron **still** has an
energy equal to

DOES THE INERTIA
OF A BODY DEPEND ON ITS
ENERGY CONTENT?

Albert's argument in this paper isn't a **proof.**

You can't prove a definition. All you can do is show that is makes sense.

So without driving ourselves crazy with more formulas, here's what Big Al is driving at:

1 the old definition of work ($W = Fd$), combined with

2 the new fact, nothing can go faster than the speed of light expressed by $F = \dfrac{ma}{(1 - V^2/c^2)^{3/2}}$ means that

3 the work goes into making the body heavier. Therefore

4 work adds to the inertia of a body and by implication inertia has energy and to make it CONCRETE . . .

5 the relationship between energy and inertia is $E = mc^2$

6 But remember . . . **nobody** really knows what inertia is — or why objects have it in the first place!

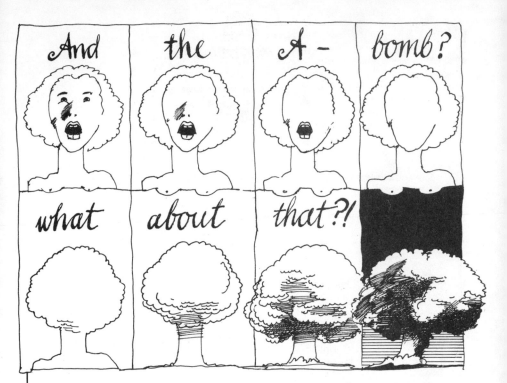

Albert just argued that energy has inertia and inertia has energy.

He didn't say anything about how to get the energy out in the first place.

$E = MC^2$ is not (as some folks think) the formula for the A-bomb. Remember, Albert proposed relativity in 1905. The A-bomb project began in 1939. Nuclear physics was developed by other scientists, like Joliot Curie, Enrico Fermi and Leo Szilard.

Szilard in 1934 came up with the idea of the "chain reaction" release of atomic energy.

Szilard wrote a famous letter, 2 August 1939, to President Roosevelt, which Einstein signed. Roughly, this letter said: Nuclear energy is here. Scientists in Nazi Germany are also working on it. Plainly, it is a decisive strategic weapon. The President must decide what to do about it.

Later, after the A-bomb was dropped on Hiroshima, Albert said: "If I knew they were going to do this, I would have become a shoemaker!"

Al, here's your Nobel Prize money.

Send it to my ex-wife Mileva, with my love.

Hmmm, gravity and electricity must be related somehow.....

Albert made other fundamental contributions to physics. His general theory of relativity (1916) was a new relativistic theory of gravitation which replaced Newton's old theory.

And Albert was a central figure in the debates raging round the quantum theory — a new theory of the electron.

Albert's materialist questioning attitude had encouraged a younger generation of research-workers to overthrow even **more** of classical Newtonian physics.

These researchers went so far as to throw out the rules of cause-and-effect. (Essentially, they said you couldn't know for sure where an electron would go when you hit it. All you could say was where it "probably" would go!)

Albert didn't approve of this at all.

Niels Bohr,
Danish physicist
and founder of the
"Copenhagen School"
of quantum theory.

Up to his death in 1955 Einstein was active, opposing McCarthyism, working with Bertrand Russell on disarmament, and still worrying about how to unify electricity and gravity. It may still be done!

Albert was a radical and a Jew. He never lost his political perspective and his consciousness of being a member of an oppressed ethnic minority.

This statement on socialism, part of a longer analysis, appeared in the U.S. magazine Monthly Review in 1949 . . .

WHY SOCIALISM?

The situation prevailing in an economy based on the private ownership of capital is characterized by two main principles: first, means of production (capital) are privately owned and the owners dispose of them as they see fit; second, the labor contract is free. Of course, there is no such thing as a *pure* capitalist society in this sense. In particular it should be noted that the workers, through long and bitter political struggles, have succeeded in securing a somewhat improved form of the of the 'free labor contract' for certain categories of workers. But taken as a whole, the present day economy does not differ much from 'pure' capitalism.

Production is carried on for profit, not for use. There is no provision that all those able and willing to work will always be in a position to find employment; an 'army of unemployed' always exists. The worker is always in fear of losing his job. Technological progress frequently results in more unemployment rather than easing the burden of work for all. The profit motive, in conjunction with competition among capitalists, is responsible for an instability in the accumulation and utilization of capital which leads to increasingly severe depressions. Unlimited competition leads to a huge waste of labor and to a crippling of the social consciousness of individuals.

This crippling of individuals I consider the worst evil of capitalism. Our whole educational system suffers from this evil. An exaggerated competitive attitude is inculcated into the student who is trained to worship acquisitive success as a preparation for his future career.

I am convinced that there is only *one* way to eliminate these grave evils, namely through the establishment of a socialist economy, accompanied by an educational system which would be oriented toward social goals.

ACKNOWLEDGEMENTS

No book is ever made by one person — or even two. Rius created the Beginner form. Members of Science for the People, of the British Society for Social Responsibility in Science and Radical Science Journal are developing a critique of science in capitalist society. Books and pictures in the British Museum Library provided enjoyable and instructive hours. Steve Bernow, Paul Raskin, Ted Werntz and Altheia Jones-Lecointe are activists with whom I shared a common experience of scientific work and education. Dick Leigh read the manuscript through for errors in both physics and politics. Andrew Friend and Andy Metcalf gave their time to learn far more about relativity than they ever really expected to, while suggesting improvements in structure and content. Gillian Slovo raised some fundamental questions about the role of math in describing physical phenomena, as well as providing friendly support in the sticky patches. Sara Baerwald and Ed Lebar put up with me for four months, while Bill Monaghan ran interference. Finally, Susie Orbach kept reminding me that I was having a good time despite all outward appearances to the contrary. It never would have got done without her.

P.S. And my thanks to Lo.

BIBLIOGRAPHY AND SUGGESTIONS FOR FURTHER READING

MAGAZINES

You can't get much from the straight press. SCIENCE magazine in the U.S. has some decent muckraking journalism, and NATURE in the U.K. runs reasonably critical reviews of current research activity. But it's heavy going and shot through with the conventional view of science as a neutral activity propelled along the paths of logic by the force of special genius.

Five magazines, started as part of the political ferment of the 1960s, express alternative views:

SCIENCE FOR THE PEOPLE (bimonthly), 897 Main Street, Cambridge, Mass. USA 02139.

IN ENGLAND:

SCIENCE FOR PEOPLE (quarterly), 9 Poland Street, London W1.
RADICAL SCIENCE JOURNAL (yearly), 9 Poland Street, London W1.
UNDERCURRENTS (bimonthly), 27 Clerkenwell Close, London EC1.
SCIENCE BULLETIN (quarterly), 27 Bedford Street, London WC2.

BOOKS ABOUT SCIENCE AND SOCIETY

The best things were produced in the 1930s by a group of radical and Communist scientists in Britain. These should be read along with Gary Werskey's constructive critique, THE VISIBLE COLLEGE, Allan Lane, London 1978.

J. D. Bernal, EXTENSION OF MAN, Palladin, London 1964/MIT, Cambridge, Mass. 1972. A history of physics. SCIENCE AND INDUSTRY in the 19th CENTURY, 2nd ed., Routledge and Kegan Paul, London 1970/Indiana Univ. Press, Bloomington 1970. Shows how science follows technology. SCIENCE IN HISTORY, 3rd ed., C. A. Watts, London/MIT, Cambridge, Mass. 1969. *"Often enough the ideas which the statesmen and the divines think they have taken from the latest phase of scientific thought are just the ideas of their class and time reflected in the minds of the scientists subjected to the same social influences."* N. Bukharin, ed., SCIENCE AT THE CROSS ROADS: Papers Presented to the International Congress of the History of Science and Technology, 29 June — 3 July 1931, by Delegates of the U.S.S.R., Kniga (England) Ltd., London WC2 1931. The paper by Boris Hessen on NEWTON'S PRINCIPIA is a classic analysis. And here is Bukharin on the neutrality of science: *"The idea of science for science's sake is naive. It confuses the subjective passions of the professional scientist, working in a system of profound division of labor, in conditions of a disjointed society in which individual social functions are crystallized in a diversity of types, psychologies, passions with the*

objective social role of this kind of activity as an activity of vast practical importance." C. Caudwell, THE CRISIS IN PHYSICS, John Lane, London 1939. An important attempt to criticize the content of physical theories.
Lancelot Hogben, MATHEMATICS FOR THE MILLIONS, George Allan Ltd., London 1937/W. W. Norton, New York 1968.
The Open University Course, SCIENCE AND THE RISE OF TECHNOLOGY SINCE 1900. Open University, Bletchley, Bucks., U.K. A useful introduction with lots of good pictures.
HARD TIMES: EMPLOYMENT, UNEMPLOYMENT AND PROFESSIONALISM IN THE SCIENCES. A Science for the People Pamphlet, London 1973, 52pp. A description and analysis of present conditions of academic scientific work.

BOOKS ABOUT EINSTEIN

There are a few standard works by people who knew Einstein, each of which has strengths. But a good comprehensive picture has yet to be drawn.
A. Einstein, IDEAS AND OPINIONS, Souvenir Press, London/Dell, New York, 1973. A collection of Einstein's ideas about politics, science and religion.
Phillip Frank, EINSTEIN: HIS LIFE AND TIMES, Jonathan Cape, London 1948/Alfred Knopf, New York 1953.
Anton Reiser, ALBERT EINSTEIN: A BIOGRAPHICAL PORTRAIT, Butterworth, London 1931.

P. A. Schlipp, ed., ALBERT EINSTEIN, PHILOSOPHER — SCIENTIST, Library of Living Philosophers, Evanston, Ill. 1949. The closest thing to an autobiography.
Carl Seelig, ALBERT EINSTEIN: A DOCUMENTARY BIOGRAPHY, Staples Press Ltd., London 1956.

MORE RECENT WORKS ARE:

R. W. Clark, EINSTEIN, THE LIFE AND TIMES, Hodder and Stoughton, London/Avon, New York 1971. Lots of facts ruined by the author's thinly veiled hostility to Einstein's politics.
Lewis S. Feuer, EINSTEIN AND THE GENERATIONS OF SCIENCE, Basic Books, New York 1974. The first book to confront the politics of the times in any depth. But Feuer's bitter opposition to the student rebellions of the 1960s has produced an odd and unworkable theory of generational conflict as the moving force in science.
C. P. Snow, VARIETY OF MEN, Scribners, New York 1971. A nice portrait from an elitist vantage point.

BOOKS ABOUT RELATIVITY

There are thousands. The trick is to find ones that seem to make sense and stick with them. Working from three or four at once can be helpful. But there's no substitute for talking the ideas over with friends.
I have based my own presentation on Einstein's 1905 paper and on his popular book which closely follows the outline of the 1905 paper.

A. Einstein et al., THE PRINCIPLE OF RELATIVITY, Dover, New York 1952. A collection of papers on special and general relativity.
A. Einstein, RELATIVITY, Methuen, London 1916/Crown, New York 1961.
J. Bernstein, EINSTEIN, Fontana, Collins Glasgow/New York 1973. An overview of all of Einstein's work.
L. Landau and Y. Rumer, WHAT IS THE THEORY OF RELATIVITY?, MIR Publishers, Moscow 1970/Basic Books, New York 1971. A popular Soviet account.
If you're not put off by the math, textbooks can be quite helpful because the accounts are nice and brief. Here are another two somewhat advanced but useful books:
THE FEYNMAN LECTURES ON PHYSICS, volume 1, Addison Wesley, London/Reading, Mass. 1963. Chapters 15-16 contain Feynman's comments about relativity.
L. Landau and E. Lifschitz, THE CLASSICAL THEORY OF FIELDS, Addison Wesley, London/Reading, Mass. 1951. A graduate level text, but pages 1-4 are an exceptionally clear outline of the theory.

ADDITIONAL BACKGROUND READING

(∗) advanced texts

W. Abendroth, A SHORT HISTORY OF THE EUROPEAN WORKING CLASS, New Left Books, London 1965/Monthly Review, New York 1972.
E. Anderson, HAMMER OR ANVIL: The Story of the German Working Class Movement, Victor Gollancz, London 1945/Oriole Editions, New York 1973.

E. T. Bell, THE DEVELOPMENT OF MATHEMATICS, McGraw-Hill, London/New York, 1940.

J. D. Bernal, THE SOCIAL FUNCTION OF SCIENCE, MIT Press, 1967.

G. Barraclough, ORIGINS OF MODERN GERMANY, Blackwell, Oxford 1947/Putnam, New York 1973.

C. B. Boyer, A HISTORY OF MATHEMATICS, Wiley, London/New York, 1968.

R. Courant and H. Robbins, WHAT IS MATHEMATICS?, Oxford Univ. Press, London/New York, 1941.

H. Cuny, ALBERT EINSTEIN, Souvenir Press, Paris 1961.

P. Dunsheath, A HISTORY OF ELECTRICAL ENGINEERING, Faber and Faber, London 1962/MIT, 1969.

A. Einstein, LETTRES A MAURICE SOLOVINE, Gauthier Villars, Paris 1956.

ENCYCLOPEDIA JUDAICA, Macmillan, Jerusalem 1971.

*J. D. Jackson, CLASSICAL ELECTRODYNAMICS, Wiley, London/New York 1972.

H. G. Garbedian, ALBERT EINSTEIN MAKER OF UNIVERSES, Funk and Wagnalls, New York 1939.

C. C. Gillespie, ed., DICTIONARY OF SCIENTIFIC BIOGRAPHY, Scribners, New York 1972.

B. Hoffman, ALBERT EINSTEIN, CREATOR AND REBEL, Hart Davis, London/Viking Press, New York 1972.

*M. Jammer, CONCEPTS OF MASS, Harper Torchbooks, New York 1964.

*F. A. Jenkins and H. E. White, FUNDAMENTALS OF OPTICS, McGraw-Hill, London/New York 1965.

D. K. C. MacDonald, FARADAY, MAXWELL AND KELVIN, Anchor Books, New York 1964.

P. W. Massing, REHEARSAL FOR DESTRUCTION: A Study of Political Anti-Semitism, Harpers, New York 1949.

*W. D. Niven ed., THE SCIENTIFIC PAPERS OF JAMES CLERK MAXWELL, Dover, New York 1965.

V. I. Lenin, IMPERIALISM, THE HIGHEST STAGE OF CAPITALISM, Foreign Languages Press, Peking. 1965.

*A. O'Rahilly, ELECTROMAGNETICS, Longmans Green and Co., London 1938.

*W. K. H. Panofsky and M. Phillips, CLASSICAL ELECTRICITY AND MAGNETISM, Addison Wesley, London/Reading, Mass. 1955.

E. J. Passant, A SHORT HISTORY OF GERMANY 1815-1945, Cambridge Univ. Press, Cambridge/New York, 1959.

*W. Pauli, THE THEORY OF RELATIVITY, Pergamon, New York 1921.

P. G. J. Pulzer, THE RISE OF POLITICAL ANTI-SEMITISM IN GERMANY AND AUSTRIA, Wiley, London/New York 1964.

E. Sagarra, A SOCIAL HISTORY OF GERMANY 1648-1914, Methuen, London 1977.

H. Schwab, JEWISH RURAL COMMUNITIES IN GERMANY, Cooper Book Co., London 1956.

*A. Sommerfeld, ELECTRODYNAMICS, Academic Press, London/New York 1952.

F. Stern, GOLD AND IRON, George Allen and Unwin, London/Knopf, New York 1977.

D. Struik, A CONCISE HISTORY OF MATHEMATICS, 3rd ed., Dover, NY 1967.

*E. F. Taylor and J. A. Wheeler, SPACETIME PHYSICS, W. H. Freeman, London/San Francisco 1963.

*S. Weinberg, GRAVITATION AND COSMOLOGY, Wiley, London/New York 1972.